Flexible Ridesharing

Volker Handke · Helga Jonuschat

Flexible Ridesharing

New Opportunities and Service Concepts for Sustainable Mobility

Bundesministerium für Verkehr, Bau und Stadtentwicklung

Fraport

Springer

Volker Handke
Institut für Zukunftsstudien und
Technologiebewertung GmbH (IZT)
Berlin
Germany

Helga Jonuschat
Innovationszentrum für Mobilität und
gesellschaftlichen Wandel GmbH
Berlin
Germany

ISBN 978-3-642-11344-4 ISBN 978-3-642-11345-1 (eBook)
DOI 10.1007/978-3-642-11345-1
Springer Heidelberg New York Dordrecht London

Library of Congress Control Number: 2012949065

Printed on acid-free paper

Springer is part of Springer Science+Business Media (www.springer.com)

Acknowledgments

We would like to thank our official project partners Lufthansa and Fraport AG Frankfurt Airport Services Worldwide. Further Klaus Hohmann from e-Notions as the initiator of the CARRIVA Ride Club. Moreover, we would like to thank the Federal Ministry of Transport, Building and Urban Development, respectively the TÜV Rheinland for supporting the project. Finally, we show appreciation to all service providers and users that have given us valuable information on ridesharing potentials as a sustainable alternative to single car use.

Contents

Chapter 1
Introduction

Individual mobility is one of the most important needs of modern society and the basis for private, public and economic life. However, transport also entails severe environmental and social burdens that do not respond to the current global guideline of sustainable development. By being one of the major sorces of CO_2 emissions, transport is a prominent driver for climate change. Moreover, individual car traffic is responsible for nearly a third of the total energy consumption.

Transport accounts for 29 % of the German end energy use (UBA 2011). In Europe, road traffic has produced 95 % of all greenhouse gas emissions, which has been caused by passenger light-duty vehicles, i.e. private transport, to a share of 45 % (EEA 2012). Moreover, accounting for a nearly quarter of energy related carbon dioxide emissions, transport is one of the main causes for climate change. In 2009, road transport has caused a fifth of the total CO_2 emissions (UBA 2011) and half caused half of all NO_x emissions in Germany (EEA 2012).

It is therefore obvious that we urgently need efficient means to reduce car traffic in order to mitigate climate change effects. However, changing to more sustainable transport modes like walking or cycling is restricted to short distances and predominantly leisure time. Thus, carpooling[1] represents a promising alternative to reduce car traffic by providing nearly the same comfort as journeys in single occupation vehicle (SOV). This is particularly interesting for commuters that both demand a certain security of travel options and comfortable travelling to or from work. Due to their relatively constant travel patterns on the same routes, commuters also represent the most interesting target group for carpooling services that need a critical mass of users to be viable.

During the last decades, different carpooling services spread across Europe and the U.S., both for long and for short distances. Most of them are based on web services connecting drivers to passengers in order to reduce individual car traffic in

[1] Unlike the British definition, we use the term "carpooling" in terms of ridesharing, offered by a car owner to a passenger, and carsharing in terms of using a common, mostly commercial car pool.

V. Handke and H. Jonuschat, *Flexible Ridesharing*,
DOI: 10.1007/978-3-642-11345-1_1, © Springer-Verlag Berlin Heidelberg 2013

general. With regard to an average number of 1.4 passengers per ride (infas 2010), carpooling could technically reduce individual car traffic by more than a half.

Carpooling services are highly dependent on the users' initiative, since drivers have to announce their rides on a website or by phone and passengers need to get active to search for a ride. During the last years, many companies have developed, general, but also customised databases and software for carpooling services that make it easy for service providers to establish their own offer without any great investments or manpower. However, most of the services do not expand their service offer by the mere existence of a website, and since most services are for free providers simply often do not know the exact number of rides initiated by their service. Actually, there is a relatively high number of "data graveyards" among carpooling websites, although particularly in areas with congestion problems the willing to carpool is relatively high.

But how can we close the gap between offer and demand? First of all, we have to understand the mobility patterns and preferences of both drivers and passengers in order to develop a carpool-friendly solution. Thus, the service has to refer to the advantages of carpooling compared to other mobility options. For drivers the main advantage is to share the costs of the ride, while for passengers carpooling must be more comfortable than public transport by about the same costs. Only if a carpooling service responds to these preconditions, it has a chance to survive and to generate at least moderate profits. A strict orientation towards customer demands is vital for carpooling services! This is particularly important with regard to the high initiative of both drivers and passengers, representing the foundation of each internet based carpooling service.

Second, only few carpooling services provide instant or flexible ridesharing that allows matching drivers and passengers in real time. This is no problem for nonrecurring long-distance rides that are planned in advance. However, main transport problems like daily congestions or urban emissions are caused by short-distance rides, notably commuting. However, a commuter needs a certain security of getting to work on time on the one hand, and flexibility of getting back home on the other hand. Therefore, a carpooling service for commuters has to offer flexible travel options in order to be a success. Though, the two main preconditions for a flexible ridesharing service is a high number of users and a real-time procuration of rides, and there are only few promising services that offer both.

In Germany, CARRIVA (www.carriva.org) is actually the only carpooling service that offers real-time ridesharing explicitly for commuters. This book is the result of our research accompanying the introduction of CARRIVA at the Frankfurt Airport, which is the main flight hub and one of the biggest employers in Germany. The research project was funded by the German Federal Ministry of Transport, Building and Urban Development, and supported by the Fraport AG, the owner and operator of the Frankfurt airport. Project partner Deutsche Luifthansa AG is one of the leading airlines in Germany.

Parallel to the commercial implementation of the service, we could analyse the commuting behaviour of the CARRIVA users, their acceptance of carpooling in general and their motivation to use real time ridesharing services. Our findings are not only useful for potential real-time ridesharing service providers, but also for transport researchers that could revive the idea of carpooling in the light of new technologies offering solutions to previous user acceptance limits.

The report starts with a general overview on mobility and commuting (Chap. 2), followed by a description of carpooling frameworks and a selection of international service offers (Chap. 3). In Chap. 4, we introduce our case study CARRIVA as introduction for the presentation of the research results according to the respective project phases (development phase in Chap. 5, test phase in Chap. 6 and outlook on further activities in Chap. 7). We conclude our report with an estimation of potential sustainability effects of carpooling (Chap. 8).

Chapter 2
Mobility

Mobility is a central concept in transport research, representing generally the need to move in space. Therefore, mobility includes all concepts of transport and traffic, but also refers to any form of movement, e.g. to social mobility in terms of changing a social status or to residential mobility in terms of changing the residence. In his book 'Sociology beyond Society', Urry (2000) defines the following forms of mobility (Urry 2000):

- "Bodily mobility" or personal mobility, of people using cars, planes, trains etc.,
- "Object mobility" of, e.g. waste, money, goods,
- "Imaginary mobility", e.g. by watching tv or movies,
- "Virtual mobility", particularly via the Internet.

He believes that all these forms of mobility are a basis for social life, creating and shaping cultural identities (Urry 2000). With regards to carpooling, bodily mobility is the central category. In the following chapters, we will therefore concentrate on definitions and frameworks of bodily mobility.

2.1 The Social Dimension of Personal Mobility

Analyzing mobility demands an interdisciplinary approach, since technical and spatial frameworks, as well as social and personal factors determine mobility patterns. In spite of a multitude of approaches, most social and spatial theories agree that today, social life is remarkably dependent on and marked by mobility. Castells (1989) definition of the "space of flows" as a new spatial paradigm is a popular example for the thesis that social space today rather emerges from dynamics and mobility than from immobile and static structures.

Mobility is also often regarded as a consequence of modernization processes (e.g. Harvey 1990), which have accelerated decision processes by electronic communication by simultaneously widening spatial boundaries and reducing

V. Handke and H. Jonuschat, *Flexible Ridesharing*,

DOI: 10.1007/978-3-642-11345-1_2,

transport costs. Modernization, thus, requires and enhances mobility, inducing social relationships that increase the mobility of people, capital, machines, information, knowledge and images. Individuals have to adapt to increasing mobility in all dimensions, which leads to the fact that mobility itself becomes a precondition for modern life.

Therefore, the concept of modernization increasingly comes along with mobility, particularly with respect to globalization processes, demanding an increasing flow of people, information and goods over longer distances. Consequently, mobility becomes more and more a subject of sociological studies and theories (Urry 2000).

On the micro-level, drivers for bodily mobility are characterized by a tension between personal resources (time, vehicle, infrastructures, money) and individual objectives (needs, preferences, activities). Here, individual objectives represent the source of mobility, while personal resources determine the form of mobility. On the macro level, individual objectives form spatial structures, while personal resources are part of the transport system. However, considering the multitude of transport studies dealing with mobility patterns and cost-benefits ratios, knowledge on social drivers of mobility, e.g. emotional or symbolic decisions, is still restricted. In this context, Urry (2000, p. 27 ff.) appoints to the following social reasons for a "co-presence" that eventually induces mobility:

- Legal, economic and familial duties, like work, family meetings (weddings, Christmas, funeral etc.) visiting public institutions (court, school, hospital etc.),
- Social duties, e.g. private face-to-face meetings at the church, local associations etc.,
- Time duties such as "quality time" with the partner, family, or friends,
- Travelling and touristic drivers,
- Events, like cinema, concerts, theatre etc.

In addition, mobility is indirectly induced by external drivers such as spatial structures, communication technologies, environmental factors, energy supply and economic frameworks. All these factors and their interactions create the socio-technical system "mobility" that determines when and why, but also how physic mobility occurs.

2.2 Personal Mobility, Transport and Transport Choices

Mobility plays a crucial role for the individual, since it represents the main connection between basic living needs for shelter, work, health care, food, communication education, and leisure time activities. The concept of basic living needs has been introduced in scientific discussion by Partzsch (1965) and Ruppert/Schaffer (1969). Since the early 1970s, several scholars have specified the definition of basic living needs (e.g. Werlen 2000) as well as its use for spatial models (e.g. Bähr et al. 1992). In spite of the debate on the definition of mobility as

Fig. 2.1 Basic living needs

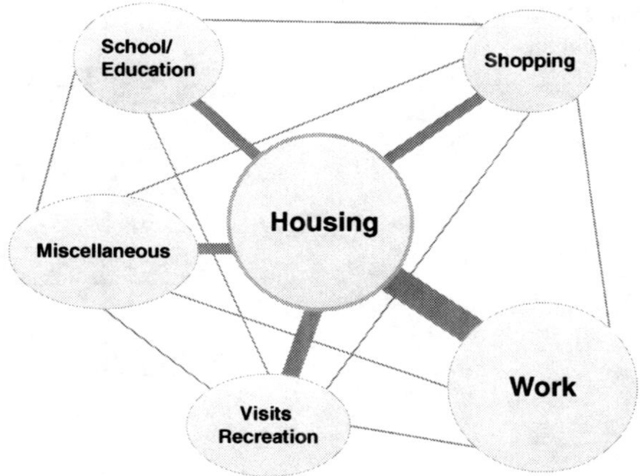

Fig. 2.2 Choice of transportation means in everyday traffic

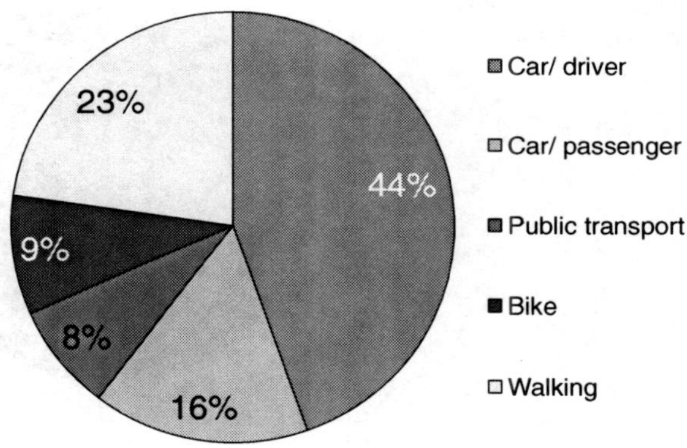

a basic living need itself, it is acknowledged that it definitely plays an important role as an enabler to fulfill other basic living needs like work, home or health care services by transport (on basis of Wikipedia 2009, GNU General Public License, see Fig. 2.1).

Regarding transport, we can distinguish modes of transport (air, road, water etc.) on the one hand, and long-distance respectively short-distance travels on the other hand. Long-distance travels are mostly associated with aviation, road, ship or rail transport and single trips, while short-distance travels tend to be regular. In Germany, two-thirds choose the car for daily trips, a quarter walks and another third uses public transport or rides a bike (infas/DLR 2010, p. 6, see Fig. 2.2).

This corresponds approximately to the share of the total of about 99 billion daily trips per means of transport (58 % car transport, 10 % public transport, 9 % walking, 23 % biking, Destatis 2006, p. 29).

Mobility connecting work and home, i.e. commuting, is one of the most relevant factors that have influenced spatial planning during the last century. Although

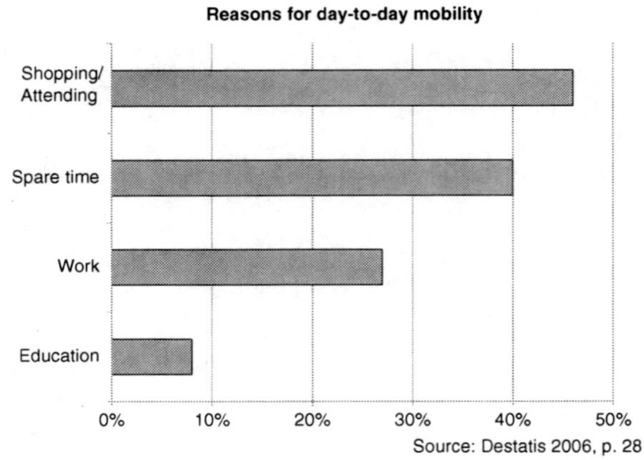

Fig. 2.3 Reasons for day-to-day mobility

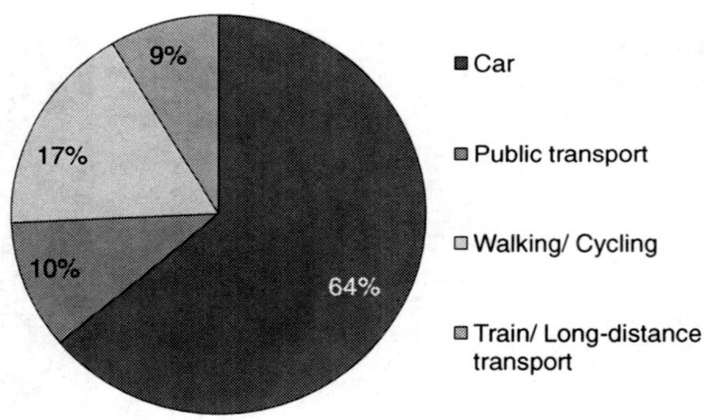

Fig. 2.4 Modal split in commuter traffic

leisure time mobility has constantly increased in the last decades, commuting is still responsible for a quarter of all private trips. The Fig. 2.3 illustrates the main reasons for daily mobility (Destatis 2006, p. 28):

The modal split for commuting has practically not changed during the last decades.

The German Federal Statistical Office states that commuting is predominantly based on car use (Destatis 2006, p. 34, see Fig. 2.4).

With regards to demographic characteristics, car use is predominantly dependent on gender as well as on income. Women cycle more (16 % of female vs. 10 % of male commuters) or walk to work (22 % of female vs. 15 % of male commuters). This is also due to the fact that men commute longer distances: 59 % of all women, but only 46 % of all men work less than 10 km away from home (Destatis 2006, p. 35).

Furthermore, transport modes are dependent on personal income. Only 48 % of people with an income of less than 500 €/month take the car for going to work—compared to 78 % of those with an income of 2,900 € or more. If the commuting distance exceeds 25 km, 86 % of people with a higher income choose the car (Destatis 2006, p. 35). Thus, main car drivers are men with a higher income.

Summarizing, about one of six daily trips are commuting trips by car. They do not only contribute to a great share to the total traffic, but also to related emissions.

2.3 Potentials for Sustainable Mobility

With about 115 million tons, passenger transport has contributed to two-thirds of all traffic related CO_2- emissions in 2003 (Destatis 2006). Between 1990 and 2003, CO_2 emissions have increased by 5 %. Since car traffic produces nearly all (98 %) of traffic related CO_2 emissions, the transport sector is crucial any sustainable development and climate protection strategy. Reducing individual transport is therefore an important step towards a reduction of greenhouse gas emissions by 21 % until 2012 compared to 1990, as defined by the Kyoto-Protocol.

Apart from greenhouse gas emissions, road traffic in general and individual car use in particular are responsible for further relevant environmental effects. Car traffic, for instance, is the main source of classical air pollutants such as NO_x, particulate matter and NMVOC, of atmospheric input of heavy metals and of organic pollutants. Cars are moreover dependent on non-renewable, mineral, metallic and fossil resources and energie, and thus responsible for related environmental effects for fuel generation and transport. In addition, car traffic is also to be blamed for noise pollution and lethal accidents.

Sustainable mobility strategies could therefore aim at motivating car users to choose low-emission transport modes like train, tramway or bike. However, only few car drivers are willing to change their mobility patterns, as the Fig. 2.5 shows (infas/DIW 2002).

According to the study “Mobility in Germany 2002”(infas/DIW 2002), about one quarter of the population (“potential public transport users”) has theoretically the potential to completely change from car to public transport, since one out of five people do not possess a car and about a quarter already uses public transport. Another third are “frequent car drivers”, stating to have an inadequate access to public transport. Consequently, only those car drivers within good access to the public transport network are an interesting target group for strategies motivating to abstain from car use.

However, this model is strongly dependent on the local or regional public transport system, resulting in strong regional variations of the model. The share of households with cars, for instance, ranges from about 50 % in Berlin or Hamburg to more than 80 % in Rhineland-Palatinate or Hesse (Destatis 2006). Furthermore, connections to public transport can also notably differ within a region. Still, the best potential shifting car traffic to low-emission traffic show social groups that have a car, but also good public transport options.

Summarizing, the group of “frequent car drivers”will not be attracted by strategies supporting public transport, but only by traffic reduction schemes that help to

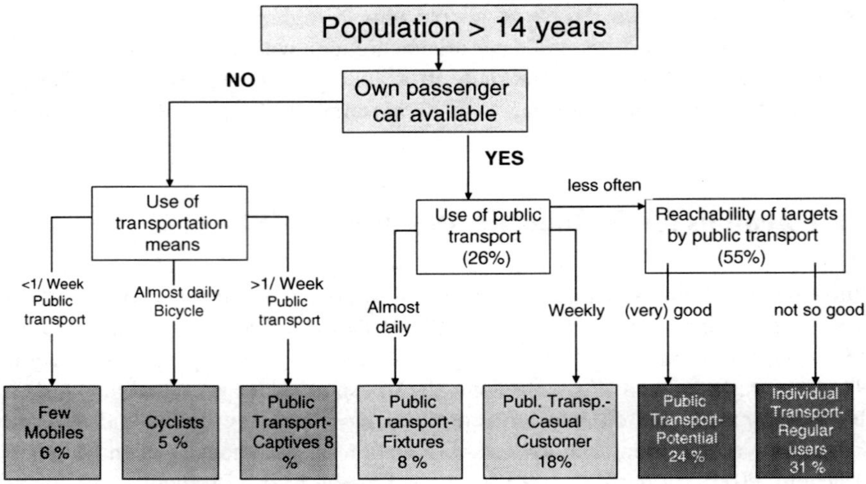

Fig. 2.5 Transportation user segments. *Source* infas/DIW 2002, p. 47

avoid single car trips, i.e. carpooling. Added to the group of car drivers with a good access to public transport, carpooling could theoretically address about half of the population over 14 years old.[1]

2.4 The Contribution of Carpooling to Sustainable Mobility

In order to reach a broad target group, a comprehensive concept of sustainable mobility has to combine advantages of individual and public transport—either by collective use of individual transport or by individualizing public transport. With regards to strategies for individualizing public transport, some initiatives currently experiment with flexible carsharing concepts that integrate individual car use as part of an overall intermodal transport concept. One of the most prominent examples is the "car2go" concept (www.car2go.com) by Daimler. Car2go-members rent any available car from the fleet, which they find in public spaces, and park it at any location after use. Similarly, Deutsche Bahn (German Railways) has initiated the "Flinkster" project that combines carsharing and public transport. In a Berlin pilot project, users can buy a "mobility card" for a monthly fee that includes unlimited travel with the public transport as well as a carsharing credit of 50 € (www.flinkster.de). Further examples are the cooperation between

[1] Although 14 year-old teenagers are not allowed to drive a car, their parents will drive them to appointments.

the Zurich public transport company and the carsharing company DenzelDrive. Consequently, carsharing schemes contribute to sustainable mobility by filling a gap of public transport systems and by reducing the need of an own car.

Carpooling, however, does not aim at substituting parts of the transport chain by public transport options, but at saving a complete trip by car. Although users of public transport are also an important target group of carpooling services, sustainability effects will only occur, if car owners abstain from using their car and instead take a lift in another car. According to the study of infas and DIW (2002), more than two-thirds of the car owners (30 % "potential public transport users" and 38 % "frequent car drivers") could potentially form a carpool, although only the "frequent car drivers" could contribute substantially to avoid traffic, since the "potential public transport users" already use public transport. This group, however, could probably abstain completely from owning a car, if they have adequate carpooling options substituting their remaining car trips.

For commuters with inadequate public transport alternatives a "guaranteed ride home" is an important incentive for carpooling, while "potential public transport users" could also use other options for a trip back home. Assuming that car drivers could give a lift for two passengers, carpooling could theoretically save 66 % of all trips. In practice, however, this potential cannot completely be exploited, since driving times and routes will always differ and therefore not lead to a perfect match. Thus, even if all car drivers are willing to carpool, personal mobility patterns times will restrict the total potential of car trip savings. These mobility patterns are dependent on spatial frameworks on the one hand and on personal preferences and mobility needs on the other hand, which determine regional carpooling potentials.

In summary, two-thirds of all car owners could theoretically contribute to save two-thirds of all car trips. The real carpooling potential, though, is dramatically reduced by local mobility structures, and only carpooling schemes that consequently adapt to these structures will success in contributing to sustainable mobility. In the next chapter, we will therefore present critical framework conditions for carpooling as well as good examples of successful carpooling services.

Chapter 3
Carpooling

In this chapter, we firstly define carpooling (Sect. 2.1) and its framework conditions, which determine the basic potential for carpooling (Sect. 2.2). Afterwards, we will give an overview on and characterize present national and international carpooling offerings (Sect. 2.3).

3.1 Definition of Carpooling

There are some inconsistencies regarding the use of the terms carpooling, car-sharing and ridesharing. In general, carsharing is used for services that provide their car pool to its customers for a monthly fee. Instead, carpooling is widely used for the practice of providing a ride to other passengers in the own car. In British English, these terms are conversely used, i.e. carsharing for sharing a ride and carpooling for using a commercial car pool, which indeed makes more sense. However, the British use of the terms is exceptional and we therefore also refer to the other use is more common, and we therefore also refer to carpooling as rides in private cars and carsharing as the use of a commercial car pool.

Apart from that, ridesharing is commonly used synonymous to carpooling, but in principle, it also includes other forms of ridesharing such as sharing taxi or bike trip. Thus, some ridesharing services such as Liftshare (cp. Sect. 4.3.10) or HOVER (cp. Sect. 4.3.21) also offer travel mates for taxis, bike tours or even walking. In our report, carpooling is consequently defined as ridesharing in a private car. Although lines are blurred, carpooling rather refers to commuters and therefore regular rides, while ridesharing is commonly used for single, long-distance rides as well. Moreover, carpooling has a connotation that refers to a more private arrangement between driver and passenger that could, but does not have to comprise a financial contribution from the passenger. As we concentrate on short-distance, regular trips to work, we will use the term carpooling instead of ridesharing throughout this book.

V. Handke and H. Jonuschat, *Flexible Ridesharing*,
DOI: 10.1007/978-3-642-11345-1_3,

The idea of carpooling was born in the 1970s, when the oil shortage caused a rise of fuel costs in the U.S.A., which motivated some people to share their rides with colleagues or neighbours. Since then, various formal and informal forms of carpooling have developed, mainly differing according to their formal character and the mode of matching. In general, we can distinguish the following three main groups of carpooling practices:

- *Informal carpooling* (slugging/casual carpooling): Informal carpooling is in particular practiced in areas that have special lanes for cars with more than one occupants (HOV = highly occupied vehicles), like in Washington D.C., Los Angeles or San Francisco. Here, commuters can be picked up near a HOV lane at a bus stop or a parking lot. Drivers take them to a central area, usually in the inner city (Burris and Winn 2006). In the case of informal carpooling, drivers generally do not take any contributions to the costs from the passenger, since the main advantage is to use the HOV lanes and therefore to be faster at the destination. Sometimes, the passengers do however pay or contribute to the fees for bridges or motorways.
- *Pre-arranged carpooling*: This is the most common form of carpooling, since internet-based services that are relatively easy to provide demand a trip planning in advance. Consequently, many different municipalities, but also commercial providers have developed pre-arrangement ridesharing systems and services during the last years in order to respond to traffic problems. These systems use internet based databases that match the drivers' and passengers' travel preferences, so that drivers and passengers can get in contact and arrange a meeting point in advance. Usually, these services expect the passenger to pay the driver a contribution for the trip. Moreover, some of pre-arrangement carpooling services also demand a small fee for the service, e.g. a monthly fee for being registered as a member or standard fee, if the matching was successful.
- *Real-time carpooling* (instant, ad-hoc or dynamic carpooling): With respect to the ongoing flexibilisation of working times, fewer and fewer employees have fixed commuting patterns. Therefore, many service providers have tried to develop services that support a real-time ride matching, e.g. by the use of dedicated parking lots, call boxes, or mobile phones. Here, the call for a ride should be instantly matched with a compatible offer in order to keep waiting times as short as possible.

This typology relates to the organisational perspective, i.e. informal carpooling is similar to private hitchhiking, while pre-arrangement or flexible carpooling is commonly associated to a ride matching service and thus, to a certain service provider who also guarantees a basic security for drivers and passengers. Due to security reasons, pre-arranged carpooling services demand a registration on a website that is often connected to social networks today (like e.g. zimride.com). Consequently, from the user's perspective, informal carpooling does indeed differ from a carpooling service that provides a more or less real-time ride matching. However, referring to flexibility as one of the main user demand factors, informal carpooling is comparable to real-time carpooling, while internet based ridesharing services need a forward planning of at least one, but usually several hours. Therefore, many researchers, service providers and organizations, e.g. the (American Transportation Research Board) use the term "flexible carpooling" for both, informal and real-time carpooling.

Summarizing, real-time carpooling can either be described as informal carpooling that is supported by a formal organisation or as pre-arranged carpooling with a very short timeframe for getting a lift respectively a passenger. It is always flexible, but needs an intervention by a local authority or a service provider to match drivers and passengers by means of ICT or on-site pick-up points.

In order to determine the flexibility of a carpooling mode, the main classification according to the basic forms of organisation (informal, pre-arranged or real-time) can be specified by further characteristics:

- *Spatial flexibility*: Real-time carpooling on certain pick-up points (e.g. a call box, a 'carpool' stand or a parking lot) and informal carpooling demand a simultaneous presence of driver and passenger on site. Using mobile or smart phones, both partners can freely determine a meeting point via internet or phone call. According services can thus relate to fixed or flexible meeting points.
- *Time flexibility*: Pre-arranged, internet-based carpooling needs time for the communication between driver and passenger, which have to arrange a place and time for meeting. However, in times of mobile communication, internet-based services can also integrate real-time communication, e.g. by interconnecting phone calls or by adapting to smartphone applications. Since on-site matching at a pick-up point could also entail waiting times for the right match, up-to-date internet based carpooling services can reach a comparable time flexibility.
- *Role flexibility*: Carpooling forms using parking lots as pick-up points as well as pre-arranged carpooling give an opportunity to negotiate who will take the role as driver. Therefore, the role flexibility is an essential factor for reducing traffic, since passengers will intentionally refrain from using their own car—instead of changing from public transport to carpooling.
- *Route flexibility*: With few exceptions, carpooling systems presume a fixed route from one starting point to a certain destination, mostly from the outskirts to the city centre in the morning and reverse in the afternoon. Particularly carpooling services using pick-up points are fixed to the routes between the city centre meeting points and parking lots in the suburbs or at highway exits. In this context, pre-arranged, internet-based services offer more route flexibility, since meeting points can vary according to the arrangements between drivers and riders.

With regards to these different dimensions of flexibility, carpooling services face the challenge of finding a balance between a constant and predictable offer and simultaneously providing spatial, time, role and route flexibility. Here, the precondition for matching a driver with one or more passengers is that their mobility patterns are as congruent as possible. In the next section, we will explain, which factors are decisive for a "perfect carpooling match".

3.2 Matching Mobility Patterns as a Precondition for Carpooling Partners

Mobility patterns have a strong influence on the success of carpooling services and can give a general hint on the potentials of avoiding traffic by the increasing use of carpooling. They differ strongly according to spatial structures, cultural influences and lifestyles, but the primary precondition of convergent routes and travel times applies to all carpooling offers around the world.

3.2.1 Route Convergence

Many American, but also Asian or African metropoles still undergo strong suburbanisation processes, which is the main cause for commuting traffic. Rush hours on certain routes visualize the daily streams of suburbanites to their work in the city centre and back. Furthermore, downtown areas also affected not only by severe congestion problems, but also by a lack of parking lots and high fees for parking, while in most parts of the world, suburban areas are insufficiently connected to an efficient public transport system. Particularly in metropolitan areas, carpooling promises to relieve burden by avoiding single occupant car trips and consequently to reduce the traffic volume as a whole. However, highway congestions do not automatically imply that all car occupants indeed have the same route.

Referring to the individual case, a convergence of routes is given, if the driver and the passenger have the same point of departure and the same destination. In case of accumulated working places such as in downtown areas, bigger industrial complexes or commercial centres, there is a high probability that drivers and passengers share the same destination. Moreover, areas with a high concentration of workplaces are usually connected to public transport. Thus, the great challenge is to match the starting point, i.e. the individual's home.

Drivers can pick up passengers and let them get off on the route. So, the point of departure respectively the destination of the passenger does not have to be the same as the driver's, but must be on the same route. Here, passengers will be probably willing to walk or cycle to the meeting point, but in case of taking the public transport or the own car to the pick-up point, advantages of carpooling must be very obvious. Here, main incentives for a passenger are for example high bridge or highway tolls, HOV lanes, a lack of (reasonably priced) parking lots at the destination or long travel distances. All of these incentives apply to American and some Asian metropolitan areas, but usually not to European cities.

Thus, the main question for the route convergence is to identify passengers that share the same route as the driver. However, there will often not be a perfect route convergence and therefore, it is important to determine the ranges that both passengers and drivers still accept. While for the passenger it is important to reach the meeting point easily, preferably by walking or cycling, the willingness of making a detour is decisive for the route convergence on part of the drivers.

Apart from the individual case, some areas feature better characteristics for a route convergence than others. The following socio-spatial factors can therefore give a hint, if driver and passenger in direction take the same route:

- *Central destination*: Many commuters share the same destination, like city centres, industrial areas, or airports traffic junctions. Therefore, a concentration of working places is a main factor for route convergence, since at least their destination coincides.
- *Central starting point*: In comparison to the destination, i.e. a concentration of workplaces, the individual's residences are usually widespread. However, starting point concentrations could also occur, e.g. in condensed residential areas or at public transport nodes.
- *Traffic corridors*: Apart from the starting point and the destination, a route convergence can occur on main routes. Such traffic corridors are particularly highways with a high traffic volume.
- *Bottlenecks*: Some routes pass through certain bottlenecks like bridges or ferry terminals. Here, ramified routes merge, which causes an increasing traffic volume. Particularly, if bottlenecks are connected to fees like in the case of the Golden Gate Bridge in San Francisco, some highways or on ferries, it could be attractive for drivers to share the fees with passengers.

The convergence of routes for a certain area can be estimated on the basis traffic statistics, but also on basis of a spatial evaluation, including an analysis of the transport infrastructure, as well as data on concentrations of workplaces or inhabitants. Concerning route convergence, complex socio-spatial patterns and dispersed transport structures can have an ambivalent effect that could either result in a dense network of carpooling options or rather in widespread, divergent route pattern.

However, even if the given spatial and transport structures are favourable for carpooling, drivers and riders will not match, if they do not travel at the same time. Thus, in addition to route convergence, time convergence of the trips is the second precondition for carpooling.

3.2.2 Time Convergence

If the travel time does not coincide, drivers and riders cannot even share a ride, if they live in the same house and work at the same place. Thus, carpooling is dependent on the fact that drivers and passengers wish to travel at the same time. Though, we have to consider that time convergence offers a certain tolerance, since usually one or both carpooling partners do not have to arrive at their work exactly in time. Due to an increasing flexibility of working times, time ranges for arrival at and departure from work can comprise more up to 2 h. Thus, while the

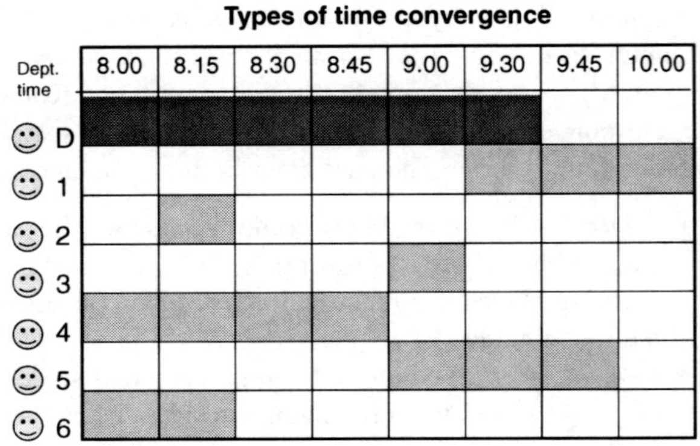

Fig. 3.1 Types of time convergence

route convergence is determined by the road from home to workplace, time convergence depends on individual, mostly even dynamic parameters. Consequently, it is vital to determine and consider the riders time tolerance for travelling in carpooling service concepts in order to increase the chance of matches. Yet, most carpooling services actually do not take account of flexible travel times, but demand carpoolers either to declare a certain departure time on a website or to be personally on site at a pick-up point, waiting for the next lift. Figure 3.1 illustrates different types of time convergence (markings show indicated departure times):

This example shows that in theory, the driver "D" shows a theoretical time convergence with 6 riders. His actual departure time, though, determines the number of "real" matches: If he, for example, departs at 8.15, he could have three passengers, at 8.30 only one, and at 9.00 two, but different passengers. In general, for a matching success, it is particularly advantageous to have broad time corridors for travelling. This does, however, increase the logistical effort for the matching procedure. Therefore, usual internet databases calculate rather with exact dates, e.g. assuming that a rider or driver wants to depart at an exact time with a certain time tolerance ahead and later.

Moreover, calculating with time spans implies that each user has to be deleted from the system, if he has actually got a match. Consequently, the carpooling database has to be regularly updated in order to match flexible departure patterns. Thus, even if an internet-database intelligently reflects departure time spans, the problem is to get constantly updated data from flexible carpooling partners.

Logistical challenges are basically the reason why most existing internet-based carpooling services force users to declare regular departure times or time spans, neglecting the fact that in reality, they are much more flexible in their driving patterns. Therefore, common internet-based carpooling services do not sufficiently address potential carpoolers who cannot or wish not to determine an exact departure time.

With regard to the frameworks for time convergence, if there are temporal concentrations of travels, which are particularly visible in case of rush hours. A higher flexibility of travel times is furthermore probable, if the given working environment is marked by flexible time structures.

Summarising, convergence of time is harder to determine than route convergence, because the commuters' travel times are in most cases flexible to a certain degree. Moreover, riders and drivers will probably adapt their usual mobility patterns in case of a carpooling option. If there is a general route convergence, matching options increase with the time flexibility of travels.

3.2.3 Integrating Flexibility in Carpooling Service Concepts

For matching two carpooling partners, it is a necessary condition that their trips are convergent with regards to the route and the time. While route convergence can exactly be determined as the road from home to work, time convergence is dependent on the individual's mobility patterns and preferences.

Still, commuter routes can be flexible to a certain degree as well, e.g. when drivers make a detour in order to pick-up a passenger or when a passenger walk, cycle or drive to a certain pick-up point. In contrast, travel times are not completely flexible, but span within a certain period, which is, though, dependent on the individual's commuting patterns. Here, carpoolers could probably adapt considerably to other carpoolers in order to get a lift or a passenger. For a real-time carpooling service, it is therefore vital to optimize the chance for a match by using intelligent logistics that respond to the carpoolers' flexible time patterns and avoid mismatches.

An obvious solution is to allow drivers and passengers to directly negotiate a convenient meeting point and departure time, either via a personal presence at a pick-up point, a phone call or instant messaging. Actually, informal carpooling and carpooling lots are based on this principle: Riders and drivers use a carpooling pick-up point to wait for a lift or a passenger and then determine a convenient drop-off destination. These on-the-spot carpooling practices are strongly connected to transport infrastructures and regulations, which is proved by the fact that most informal and formal carpooling pick-up points are situated along HOV lanes or toll roads.

While in the case of on-the-spot-carpooling practices, passengers and drivers meet on the basis of a rough common departure time and a common starting point, internet-based carpooling services potentially allow carpoolers to even freely negotiate time and place of the departure. Yet, most internet-based carpooling services do not use this opportunity adequately, but force the users to set an exact departure time and place. This is basically due to the fact that simple databases calculate with exact times and routes, while integrating time spans and flexible routes demands more complex logistic systems. Apart from that, only few services make use of further incentives such as parking privileges or carpooling rewards like gifts, shopping or fuel vouchers. The Chap. 4 gives an overview on a selection of different carpooling service concepts that are practised around the world.

3.3 Overview on Carpooling Practices Across the World

Within our research, we have identified over hundred carpooling-projects and service worldwide (see Sect. 2.3). The research has concentrated on initiatives in Europe, USA, Canada and New Zealand, since in these regions, a broad range of carpooling services were developed, tested and evaluated as pilot projects or provided at the market. Particularly the USA and Canada have a long tradition of carpooling, which was resurrected during the oil embargo in the 1970s. In both countries, strong suburbanisation processes, a high dependence on car transport with a simultaneous lack of efficient public transport, and concentrations of workplaces in central downtown areas have been and still are a good basis for a broad acceptance of carpooling.

In Europe and New Zeeland, the idea of organising carpooling first came up in the mid 1980s, merely as an idea imported from America. In this context, public–private-partnerships have tried to develop carpooling offers in order to check out the acceptance for carpooling outside North America. However, most of these first initiatives did not survive the pilot project status, since efforts for organisation and marketing were not in relation to small number of interested participants.

Due to the diffusion of the internet during the 1990s and the spread of simple software tools for databases during the last years, carpooling services became increasingly popular in all parts of the world. Nowadays, smartphones promise to reinvent the idea of flexible carpooling by using of mobile internet and GPS applications. Today, a large number of internet-based carpooling or ride sharing services exist on each continent. In spite of the new technical opportunities, real-time carpooling services meeting commuter demands are still an exception.

In our overview on carpooling initiatives, we have characterised certain types of organising carpooling in order to demonstrate all the different approaches to introduce or support carpooling practices. This typology is based on an analysis of the organisational structure, socio-geographical framework, the service offer, technical parameters etc. of present or former carpooling services. Our research is mainly based on information on the service's webpages as well as on scientific studies and presentations of the service. In some cases, we have contacted the initiatives' representatives by phone or e-mail in order to get further information, particularly on usage data. Furthermore, we have registered as members and have checked available lifts. On this basis, we have developed typical categories of carpooling initiatives, which are each described by an exemplary service portrait in the Chap. 4.

Table 3.1 gives an overview on the carpooling practices that we have selected as typical examples. In the following table, we have classified all practices according to their character as informal, pre-arrangement, or real-time carpooling practice (cp. Sect. 2.1):

On the global level, internet-based, pre-arrangement services are the dominating carpooling practice. With the exception of some services joining carpooling partners on-site at dedicated pick-up points, most real-time carpooling services additionally use an internet-database for pre-arranged trips. Both informal examples from the U.S.A. and Indonesia resemble to on-site real-time carpooling initiatives, but are,

Table 3.1 Carpooling initiatives across the world

	Informal carpooling	Pre-arrangement carpooling services	Real-time, carpooling services
M21 FahrPlus (Germany)		X	X
Frauenfahrgemeinschaft.de (Germany)		X	
Mitfahrgelegenheit.de (Germany)		X	
Saarland-Mitfahren/parking lots (Germany)		X	X
Move Bremen Pilotprojekt (Germany)		X	X
Berufspendler Netz Sachsen (Germany)		X	
City Netz (Germany)		X	
Carlos (Switzerland)			X
Easy Rider (Netherlands)		X	X
Covoiturage (France)		X	X
Liftshare.com ltd (Great Britain)		X	
DriJo (Europe)		X	
Carpool.ca (Canada)		X	X
Ottawa Carpool parking lots (Canada)			X
CommuteSmart (USA)		X	
CommuteChoice Ride Sharing (USA)		X	
GoLoco (USA)		X	
Jayride (?)		X	
Zimride (?)		X	
Slug-Lines (USA)	X		
HOVER-System (USA/Neuseeland)			X
Nuride.com (?)		X	
HOV-Jockeys (Indonesia)	X		
Mega Carpool (India)		X	X

by definition, not organised by any formal institution. Our selection of carpooling examples is only a section of all existing services. However, we have tried to gather comparable offers and describe them by the most popular or most typical service.

Still, our sample of deeper analysed carpooling services reflects the real situation in the sense that there is a clear focus on pre-arrangement carpooling services, while informal and real-time carpooling is rather the exception to the rule. Our selection spans from unique services or pilot projects like the M21, DriJO or CARLOS project to widespread international platforms, e.g. Liftshare.com.

3.3.1 "M21 FahrPLUS" Pilot Project at Daimler-Chrysler (Germany)

In cooperation with the German State of Baden-Wuerttemberg, the Daimler-Chrysler AG has initiated the carpooling pilot project "M21 FahrPLUS" as part of the project "M21—new mobility services for the 21st century" in 1998. Apart from the carpooling service "FahrPLUS", the M21 project included also the testing of a short time car rental service, of a navigation system and of a teleshopping service. In Germany, "M21-FahrPLUS" was one of the first initiatives to establish a real-time carpooling service by using both internet and mobile phones. After 4 years, the project ended in 2002, although the plan was to hand over the carpooling service to a commercial provider. However, only few people have used the carpooling offer during the pilot project phase: Only 320 of a total of the 6000 Daimler-Chrysler employees in Stuttgart have been registered as "FahrPLUS"-users (Wegmann 2007).

The "FahrPLUS" system matched short-term ride sharing requests up to 1 h before departure. The users had to post their travel options on either an internet or intranet platform and were directly informed by e-mail, short message, (mobile) telephone, fax or pager on potential ride sharing matches. Each pilot project testees received a "mobility guarantee", which allowed them to use cars of the company's vehicle fleet or to use the public transport for free, if they have shared a ride for the way to work and could not find a lift back home.

The advantage of the M21 FahrPLUS system is that it is one of the first systems which adapted to dynamic departure and arrival times, and consequently to individual and flexible working patterns. However, during the pilot project phase, technical problems and an unsufficient usability performance of the communication system resulted in a poor acceptance of the service amongst the Daimler-Chrysler employees (Wegmann 2007). Thus, the project partners eventually decided not to (weiter verfolgen) their carpooling initiative.

3.3.2 Frauenfahrgemeinschaft.de: Ridesharing for Women (Germany)

The internet platform "Frauenfahrgemeinschaft.de" exclusively addresses women. The service has been launched in 2006 and aimed to offer women a safe ride avoiding the potential of sexual offences during the ride.

After registration as user, requests and offers can be posted on the service's website, and interested members can directly contact each other. The provider promises a high safety standard by a "safety team" regularly checking postings and reacting on complaints.

The unique selling point of "Frauengemeinschaft.de" is a high subjective feeling of security. The homepage is only in German, so users are restricted to German speaking countries. "Frauengemeinschaft.de" has not reached a critical mass yet to provide a broad offer on main routes. A further obstacle is the competitor "Mitfahrgelegenheit.de" (see below), which is dominating the German market at least for long-distance trips, offering the possibility for "women only" for ride sharing as well.

Target groups are obviously women, both for single and regular rides, although the provider declares that most members are regular commuters. In May 2009,

2000 women were registered at Frauenfahrgemeinschaft.de. The website structure does not offer the opportunity to get data distinguishing between drivers and passengers or data to detect the actual number of ridesharings matched by Frauenfahrgemeinschaft.de.

3.3.3 Mitfahrgelegenheit.de: Ridesharing on Long-Distance Trips (Germany)

The ridesharing service Mitfahrgelegenheit.de declares to be one of the biggest independent carpooling services in Europe. The service was established in April 2001 and has about 850,000 registered drivers and passengers. As in the last example, the service providers cannot determine the number of actual matches and shared rides as well as the proportion of drivers to passengers.

Mitfahrgelegenheit.de concentrates on single and long-distance trips. It offers the individual publication of personal information, a search for travels abroad to 45 countries, "women only" lifts, showing meeting points by Google maps, the use of an internal mailbox system as well as a posting of regular rides. Furthermore, there is an opportunity to mark so called "black sheep", i.e. carpooling partners behaving negligent or irresponsible. Paying an additional fee, users have to authenticate by their identity card and therefore profit from a higher security level.

Mitfahrgelegenheit.de was originally established in Germany, but has expanded its offer to Switzerland, Austria and Greece. Further European countries will follow in future, which is already visible as links on the service's website. Mitfahrgelegenheit.de has indeed reached passed the goal of having a critical mass of users that guarantee a constant and a broad offer. However, commuters have not been attracted by the service yet, although there is indeed a possibility exists to post regular trips within a certain region.

Using synonyms makes it hard to track a person in case of problems or of "professional drivers" who serve certain routes e.g. with vans as a business. Apart from the fact that profit-oriented drivers evade taxes and business declarations, lifts can be inconvenient for passengers, e.g. if the driver is distracted because of long driving times, lack of reliability etc.

3.3.4 Saarland-Mitfahren.de: Carpooling on Parking Lots (Deutschland)

Saarland-Mitfahren.de is an internet gateway for commuters in the German state of Saarland at the French border. It represents the only European service adapting the Northern American idea of on-site-matching at a parking lot.

The state-owned, public transportation company VGS (Verkehrsmanagement-Gesellschaft Saar mbH) is responsible both for the website and the parking lots. Saarland.Mitfahren.de cooperates with pendlerportal.de, which is a central platform for most West German ridesharing services.

On the one hand, passengers and drivers can get anonymously in contact on the website and arrange a meeting either on a carpooling parking lot or at any another meeting point. On the other hand, carpoolers can personally get into contact on site and spontaneously arrange a lift. All 38 carpooling parking lots of Saarland-Mitfahren.de are listed on their website. They are located at highway exits or important national route junctions, and parking is free of charge.

As carpoolers use the website only for the first contact and separately arrange meeting points and times, Saarland-Mitfahren.de has no information on actual rides and costs for the lifts. Providing carpooling parking lots is a unique feature of the service at least in Europe. However, there is a risk of burglary or theft on the remote and unsupervised parking lots. Therefore, on the website, owners can get recommendations, how to protect their cars from crime.

3.3.5 The MOVE Project in Bremen (Germany)

From 1995 to 1998, the Environmental Agency in Bremen has coordinated the project MOVE, which was funded by the EU (cp. Umweltsenat 1998). The goal of the project was to develop a software allowing ridematching both via internet and in person. Further project partners were the Bremen branches of two big companies (Kraft Jacobs Suchard and Beck's Brewery).

The implementation of the project was supported by various supporting measurements:

- Information and motivation campaigns at work,
- Job ticket for the public transport,
- Compilation of individual public transportation schedules,
- Parking privileges for carpoolers,
- A "guaranteed ride home" for public transport and carpool users,
- A consulting service for all mobility issues,
- Assistance in forming a carpool (e.g. potential carpoolers meetings, information on assurances and taxes, calculating a personal CO_2 balance etc.),
- Carsharing stations on the premises as well as special conditions for participation in car sharing.

The project team has developed a business model, which is based on the idea that companies pay a fee for establishing the MOVE concept. The fee depends on the company size, and ranges from 17 € (<50 members) to 600 €. For a monthly fee of 2.50 €, MOVE members get a "guaranteed ride home" allowing on monthly taxi ride of up to 25 € in case of a lack carpooling options for the ride home. Alternatively, carpoolers can use car sharing services for free.

The project report does not state the actual number of users, but states that the "critical mass" could not be reached. With regards to the critical mass, the project team assumed that 6000 persons have to use the service, providing a chance for a match of 20 % at each request. Yet, only about 2000 employees work at the companies Kraft Jacobs Suchard (today Kraft Foods) and Beck's. Thus, the project team has come to the conclusion that they have to concentrate on canvassing further participants from other companies rather than on the technical solution.

One measurement was, for instance, to integrate the works committee and further staff from the human resources department in public campaigns supporting the idea of carpooling as a sustainable mobility option. However, the works committee succeeded in preventing planned carpooling parking lots, which are a central incentive for carpoolers (Umweltsenat 1998, p. 31). With regards to this experience, the project team emphasised the need to participate the works committee at an early stage.

In the end, the project team had to admit that they have underestimated the efforts to launch of a comprehensive strategy for recruiting members, which is crucial for the success of carpooling services. Further recommendations of the project team are:

- Mobility concepts have to be consequently adapted to the location resp. the company,
- Carpoolers do not abuse a "guaranteed ride home" offer,
- While cooperation with carsharing services for "guaranteed ride home" is useful in urban areas, participants living in rural regions have to use a taxi due to the lack of car sharing stations.
- Companies prefer marketable services to subsidised initiatives that demand a strong commitment and participation—even if they have to pay for an effective "full-service"-concept,
- Companies expect to use their action for corporate carpooling offers, e.g. for eco-balances, public relations, their image etc.
- Companies will not accept complicated and complex carpooling systems, demanding adjustments, skills or practice from both the company and the user ("The simpler the service, the better the acceptance"),
- Companies do not show interest in activities in the field of mobility, if they are not supported by legal, spatial or financial incentives such as free carpooling parking lots, fiscal advantages or environmental requirements.

The results of the MOVE project are very important for developing ridesharing services, since they point out crucial success factors for carpooling services in general and company-based carpooling concepts in particular. As a publicly funded research project, the MOVE project team has published valuable information on problems and constraints that commercial providers would not reveal.

3.3.6 *Commuters' Network Saxony (Germany)*

Since 2008, the "Commuters' Network Saxony" (www.sachsen.pendlernetz.de) offers a free ridesharing service. The organisation emphasizes both savings for fuel costs for carpoolers, and image profits for companies, who establish the service as part of a corporate mobility strategy. The network cooperates with the Saxon Energy Agency (SAENA) and plans to expand its network by approaching local authorities and companies in order to disseminate their concept. The service is available throughout the whole State of Saxony, and contains subpages for the district and county level.

According an interview with SAENA, 500 users have registered for the service (status in May 2009) and that most users live in rural areas. At that time, the "Commuters' Network Saxony" has not yet reached a critical mass for a broad carpooling offer. SAENA states that irregular or inappropriate working times represent a main constraint factor for carpooling schemes, while main promoting factors are raising fuel costs as well as an environmental concerns. Therefore, cooperating with the state energy agency could enhance the trust and confidence in the service.

3.3.7 *Citynetz: Rideshare Portal for Long-Distance Trips (Germany)*

In Germany, long distance liftsharing is quite popular for decades. Since 2003, all German liftshare services were bundled in one internet portal (www.citynetz-mitfahrzentrale.de). Whereas the first liftshare services were organised completely by phone, applications and offers are posted online today. Still, in case of a match, a Citynetz employee will call the according passenger, who can either directly book the journey by transferring his bank details via phone or personally go to a local Citynetz branch. When the payment is confirmed, the passenger gets the contact details of the driver. Citynetz charges an operation fee and passengers are expected to pay 6 Ct/km to the driver.

Although Citynetz does not focus on regular short distance trips, it represents an well-established ridematching service. The main success factor for Citynetz is that ridesharing is formalised by the personal service, which to a certain degree raises trust and security.

3.3.8 *The CARLOS Carpooling Call Box (Switzerland)*

The goal of the CARLOS project was to test an innovative ridesharing system for rural areas in the Swiss region Burgdorf (near Bern) from 2002 to 2005. The core of the CARLOS concept is a call box that displays the passenger's destination and prints out a carpooling ticket for about 1,40 € (2 CHF). In case of a matching destination, the driver gets this ticket, which can be collected and exchanged at petrol stations or public transport ticket offices. Since ten tickets value about 7 € (10 CHF), half of the ticket price is reinvested in the service. The columns are moreover equipped with schedules and a button for a taxi service as well as with a telephone connection to the CARLOS control centre. Today this form of ride sharing service from the CARLOS GmbH (www.carlos.ch) is offered as "CARLOS Classic" while the offer of ride sharing services for companies is advertised as CARLOS Business.

The CARLOS project was scientifically evaluated and the report provides valuable results for the implementation and use of ridesharing services in rural areas (Artho et al. 2005). CARLOS should complement the public transportation system in remote areas. The project was flanked by various marketing activities, including a festive opening ceremony, a press conference and a poster campaign in local trains. During the second project year, another extensive marketing campaign was launched. In order to support the introduction of the service, the CARLOS team has printed and distributed leaflets, initiated cooperation, created a homepage, and disseminated the CARLOS concept by the press and cinema spots. The marketing campaign was accompanied by quite innovative ideas like a CARLOS rallye or CARLOS experiences of lead users, which were published in the local press.

During the project, CARLOS has matched averagely 1884 lifts per, which results in an average of 5.3 lifts per day. However, the intensity of lifts has been decreased from about 300 drives (=10 drives per day) in the first month to about 60 drives (=2 drives per day) at the end of the project. In the CARLOS pilot project area, 75 % of the people have heard of the service.

Initially, the main target group were teenagers and students (until the age of 24). Although only half of them knew of CARLOS, they were indeed were overrepresented as users. About 30 % of the CARLOS users were aged between 25 and 45 years and another 30 % between 54 and 64 years.

The user survey showed that a CARLOS lift was valued as an equal alternative to car use, representing a better option to public transport in general. Still, specific criteria of the CARLOS service (speed, independency, flexibility and reliability) were rated worse in comparison to the public transport or car use. CARLOS was predominantly used in times of a lack of public transport options, e.g. at night or in case of long waiting times for the bus.

Regarding mobility, flexibility seems to be an important bottleneck: Although the average waiting time of 7 min was shorter compared to other public transportation options in the rural areas, the users named as well waiting times for a lift as main weakness of the CARLOS as well as the unpredictability of the waiting period and small number of call boxes. Furthermore, users have stated that they generally have no need for CARLOS due to other travelling alternatives. Moreover, some users could not find a ride home and reported that in total, driving costs have been higher than expected. With regards to social aspects, CARLOS has generally contributed to the willingness of providing a lift (Wuppertal Institut 2004).

Concerning the routes, the highest activity has been on main traffic corridors, e.g. between the cities Burgdorf and Kirchberg as well as to the local IKEA branch.

With regards to the investments, 10 lifts per day could cover the annual costs of about 27.000 € (~20,000 €/a for employees, ~7,000 €/a for infrastructure and by 60–80 % (Wuppertal Institut 2004).

Summarizing, the CARLOS revealed valuable information on the implementation and operation of ridesharing services in rural areas. The CARLOS call box serves as information and service point, providing an innovative option of spontaneous ridesharing service. However, in spite of the positive reaction, general users reservations must be taken seriously, if CARLOS should be implemented as an reliable alternative to public transport in rural areas.

3.3.9 Easy Rider at the Amsterdam Airport Schiphol (The Netherlands)

The project "Social Traffic" aimed to test and scientifically evaluate a flexible ridematching system at the Amsterdam airport Schiphol between 2004 and 2005. The service called Easy Rider combined mobile phone and internet on basis of a technical system that was developed and provided by the company Ecolane (www.ecolane.com). All carpooling requests were handled by a call center, which connected matching partners. Lifts were directly charged by the Bridgetools database and billing system (www.bridgetools.nl).

The project was supported by a large consortium of partners from the telecommunications industry (Orange, Alcatel, map24 etc.) from social and environmental organisations (WWF, WAR Child etc.) and from the financial sector (VISA, Rabobank etc.) as well as from larger employers at the airport (KLM, Martinair, Transavia etc.). Although the project had good starting conditions characterised by a high number of supporters and a broad media campaign (e.g. radio interviews, reports in nationwide broadcasts), only 180 users have been registered at the Easy Rider service (Social Traffic 2005).

User inquiries have revealed that the awareness of Easy Rider was relatively high, and that people showed a positive attitude regarding carpooling in general and Easy Rider in particular. As in other carpooling projects, problems arose rather due to the fact that the project consortium has underestimated the efforts of introducing and adapting the service components to the users' preferences (Social Traffic 2005). A member of the project team describes in his blog, that the system was technically advanced, but the business model and the marketing concept were insufficient resulting in high costs (Hannula 2009).

The Easy Rider pilot project illustrates that a good concept and a broad consortium of supporters will fail, if the user do not perceive the advantages of the carpooling system. The project consortium thus summarized that a successful implementation of ride sharing services need additional incentives for both drivers and passengers, as well as a large marketing budget.

3.3.10 *Liftshare.com: Carpooling Service for Private and Corporate Members (Great Britain)*

Liftshare.com offers a classical internet-based ridesharing service. Founded in 1997, it was one of the first ridesharing services in Great Britain. Liftshare has originally addressed students in the Bristol area, but is nationwide operating today. In August 2011, more than 455,000 registered users and 290,000 currently active users in about 1,300 carpooling schemes have offered more than 290,000 lifts. About a third of all requests result in a contact between members. Liftshare also asks for feedback after each contact between members, and states that 45 % of the contacts actually result in a journey.

Liftshare is free for private users. The company finances the service by offering more than 1,300 ridesharing schemes e.g. to councils, companies, universities or business parks. Big clients are, for instance, BBC, Tesco, Virgin and the Environment Agency. Liftshare announces that a third of all registered lifts lead to a successful contact. Since 2000 Liftshare has achieved almost 30 business and environment awards. For companies, Liftshare offers their software BUDi in four variations: CarBUDi, TaxisBUDi, BikeBudi and WalkBUDi. The software aims to encourage colleagues not only to carpool, but also to walk and cycle together or share a taxi.

3.3.11 *Covoiturage-Aeroports: Carpooling at Paris Airports (France)*

"Covoiturage-Aeroports" (www.covoiturage-aeroports.fr) is an internet platform of the French company "Green Cove", offering commuters a ridesharing service to and from all international airports in Paris (Charles-de-Gaulle/Roissy, Orly and Le Bourget) since November 2006. In 2011, 1,787 users were registered with an offer of 1,297 lifts. Private individuals can register on Covoiturage for an annual fee of 16 €. Companies register for free, and employees can individually register on their behalf using a company related code. There are three ways to search for a matching driver or passenger:

- By an interactive map showing potential lifts,
- By the name of the city listing all lifts at one location,
- By the via-michelin route calculator that also allows finding lifts for route sections.

Users can contact a matching carpooling partner by e-mail. Of particular interest is that the service is supported by an iPhone application called "Comuto" allowing covoiturage users to register or post a lift in real time and en route.

Starting a request, users get a list with personal short profiles of all potential drivers or passengers that can be contacted directly by phone or e-mail. Apart from the posted lift and the according CO_2 savings in comparison to a single occupied ride, personal short profiles include pictures of the covoiturage member and icons indicating e.g. non-smoking cars, options of carrying a dog, interest in social contacts and

3.3.12 DriJo: Selling Rides by Internet Auction (Germany)

DriJo is a patented system to sale long distance rides by internet auction. Similar to Ebay, drivers can post lift options on the DriJo homepage, and passengers purchase them by auction. Both posting and purchasing is for free. According to an interview with the DriJo managing director, DriJo has registered 7,000 auctions as a redult of an article on the popular German IT related online magazine heise.de (Mühlbauer 2008). Albeit the interesting and playful concept, DriJo has discontinued its service due to a lack of users after 2 years.

3.3.13 Carpool.ca: Carpooling for Commuters in Calgary (Canada)

Funded by the Climate Change Action Fund, Carpool.ca has been founded in the year 2000. It is the first internet-based ridesharing service in Canada, addressing commuters particularly from rural areas around Calgary. In 2010, about 9800 user were registered at Carpool.ca. Because of its success the service has won several awards such as Canada's Energy Efficiency Award 2003 and the Minister's Environmental Award 2002.

Carpool.ca aims to support lasting ridesharing communities among commuters with and regular travel times and routes on the long-term. The municipality of Calgary is an important partner who is actively promoting this service by organising regular campaigns, e.g. by annual "ridesharing-weeks" in October or the "Go Green for St. Patrick's Day" campaign in March. In addition, "park'n'pool" lots offer the opportunity of flexible, on-the-spot ridesharing. Furthermore, carpoolers are privileged at applying for the rare municipal parking lots in the city centre (City of Calgary 2008).

Carpool.ca is a best practice example for a holistic approach of organising carpooling on the regional level, using synergies of public incentives, support of the local authorities and a user-friendly non-commercial ridesharing service.

3.3.14 Ottawa Carpool Parking Lots (Canada)

Following the carpool.ca example, the city of Ottawa created several carpool parking lots at main routes to the city centre. The service is not accompanied by an internet website and lifts have to be exclusively arranged at one of the nine carpool parking lots. However, the internet homepage OttawaRideMatch.com as well as the homepage of the City of Ottawa map all carpool parking lots in the area.

3.3.15 CommuteSmart in Southern California (USA)

CommuteSmart.info is combines five local ridematching services in Southern California:

- Metro Commute Services (www.metro.net),
- Riverside County Transportation Commission (www.rctc.org),
- SANBAG Rideshare Program (www.sanbag.ca.gov),
- Ventura County Transportation Commission (www.goventura.org),
- Orange County Transportation Authority (www.octa.net).

Users can either register online at one of these ridematching services for free or call the Commute Hotline to instantly get a lift. The latter option without a registration has the goal to attract new users. On the website, the Ride Guide service moreover provides a personal ridesharing schedule.

CommuteSmart furthermore includes an all-inclusive service for employers located in the Southern Californian counties (Los Angeles County, Orange County, Riverside County, San Bernardino County, Ventura County), who wants to install a commuter ridesharing service for their employees. The member companies guarantee a ride home for their staff. Employees were moreover encouraged to plead for an application for their company at CommuteSmart.

Marketing activities of CommuteSmart comprise:

- Events like the Rideshare-Week 2008 or the Bike-to-Work-Day,
- A monthly newsletter,
- A poster for a free download,
- Video clips for marketing the ridesharing idea within the company or privately.

With the offer of free material, elaborate marketing activities are transferred both to companies and private individuals. Finally, CommuteSmart is also connected to with further mobility offers like Park & Ride and the public transportation.

3.3.16 CommuteChoice Ride Sharing Program at Harvard University (USA)

The CommuteChoice RideSharing program (www.commuterchoice.harvard.edu) aims to relieve the tense parking situation at the American Harvard University by promoting carpooling among students and the staff. For instance, several free parking lots are especially dedicated to intensive carpoolers that offer a lift for more than 4 days a week and the parking fees are charged according to the number of passengers. Furthermore, the University offers a guaranteed ride home for carpoolers twice a year and registered ridesharers get four vouchers for parking lots per month.

The CommuteChoice Ride Sharing program also directly initiates carpooling communities by phone or e-mail. In case of a matching, all members must officially register as a carpooling community. The CommuteChoice Ride Sharing homepage furthermore refers to the GoLoco service (see next section) in order to find carpooling partners.

3.3.17 GoLoco: Social Network for Carpoolers (USA)

In 2007, Robin Chase, the founder of the carsharing service Zipcar, has founded the company GoLoco (http://goloco.org). The name refers to the reduction of CO_2 emissions ("Go low CO_2"). GoLoco is primarily a social network site, on which members can publish a private profile with photos, personal information etc. Users can find carpooling partners by creating a group or by joining an existing group.

Partly, GoLoco groups refer to a certain route (e.g. Bedford/MIT-Commuters oder Northeastern University students) or on groups of interest (e.g. Greyhound bus owners, Rock Climbers of Metro Boston, Unitarian Universalists or "People who like to swim"). The principle that ridesharers know each other at least vaguely or share the same community of interest enhances the trust between driver and passenger and thus, relieves an important barrier to ride share.

Moreover, GoLoco offers a billing system for carpoolers allowing a virtual billing of the trip. CoLoco moves a share of 0.50 $/mile per passenger from the passengers to the drivers account, charging a 10 % transaction fee. The user account can be reloaded by Paypal.

3.3.18 Jayride.co.nz: Single and Repeating Ridesharing (New Zealand)

Since 2208, Jayride (www.jayride.co.nz) is a popular internet-based ridesharing service in New Zealand. On Jayride, only drivers can post lifts, and passengers have to contact the drivers. One reason for its success is that Jayride serves remote areas with a lack of public transport alternatives and provides both long-distance rides and repeating rides for commuters. Moreover, the Jayride homepage is quite user-friendly, giving e.g. all relevant information on rides already on the start page.

On the user profiles, members can check the last visit of other members and their past rides, which have been matched by Jayride. According to this information, about 10–15 % of the currently 6,208 users (August 2011) have already been used Jayride predominantly once or twice (own calculation on basis of a random sample of 100 members), although some members are also extraordinary active. There is a constantly a high number of updated offers and memberships have increased gradually during the last years.

In August 2011, 33 Jayride members announced on–off rides and 130 members repeating lifts. With regards to the pronouncement that the service has more than 6000 members and currently offers 6,833 lifts (=50 lifts per regular driver), one can draw the conclusion that only a small number of members (here about 3 %) are active in a certain period and that not a total number of lifts, but rather the number of drivers is important for the chances to get a lift. In any case, Jayride is a good example for the fact that transparency on the actual chances for getting a lift is an important factor for enhancing trust in the service and thus binding members.

3.3.19 Zimride.com: Ridesharing with Facebook (USA)

Zimride.com claims to be the largest ridesharing service in the United States with more than 300,000 members. It was founded in 2007 with the help of a 250,000 $ seed money from a Facebook investment funds, and users need a Facebook profile to sign in and use Zimride, which should increase the credibility of the carpoolers. During the pilot phase each participating university has furthermore invested 10,000 $, in order to guarantee a free offer to students and staff.

Zimride predominantly addresses companies (e.g. Wal-Mart, CIGNA etc.) and universities (e.g. University of Michigan, UCLA, Berkeley etc.) to integrate the service in their corporate mobility scheme. In total, Zimride has 110 clients in 30 states and has already initiated trips for more than 100 million miles (Evangelista 2011). Some companies such as Facebook, the software company Inuit and PWC (PriceWaterhouseCoopers) use Zimride in all branches across the United States. Although most users are members of a university or corporate network, they can also sing into "public zimride".

Drivers and passengers can post their destination or check out the Zimride and Facebook profiles of ride-sharing matches. If they have found a potential carpooling partner, they have to book the trip and wait for the other partner to accept it. Passengers have to pay the trip in advance via Paypal, and drivers will get the contribution after the journey.

3.3.20 Slug-Lines in Washington D.C. (USA)

During the oil crisis in the 1970s, Washington D.C. commuters have spontaneously initiated an informal carpooling system along main routes with HOV lanes (exclusive lanes for "High Occupancy Vehicles"). In the morning, drivers and passengers meet on parking lots at highway exits, arranging a lift to the city centre. In the afternoon, several spots in central business districts serve as pick-up points for the ride home. Here, passengers queue in so-called "slug-lines" and wait for a lift. A similar system of informal carpooling or "slugging" is also practiced in Houston and San Francisco.

The website www.slug-lines.com supports the informal carpooling system in Washington D.C. by providing all necessary information on certain routes for registered users, such as distance, driving-time, registered drivers for this route, rush hours, connection to the public transportation, parking facilities as well as the best return option. In addition, registered users can exchange their experience or arrange a ridesharing in a forum.

The Metropolitan Washington Council of Governments supports slugging by offering registered users four guaranteed rides home per year. In case of unexpected changes of plans or emergencies, registered commuters can use a taxi or a rental car for their trip home for free.

Each day, about 6,500 people use slug-lines in Washington D.C. (Vanasse Hangen Brustlin Inc. 2006, cited by Oliphant 2008). Oliphant (2008) has interviewed 300 "sluggers" on their experience and motivation, and has found out that the main motivation for drivers was to save time and for passengers to save money. 85 % of all respondents slug both to work and back home and most sluggers work near the White House or the Pentagon.

Oliphant (2008) compiles the following success factors for informal carpooling:

- HOV lanes for at least three occupants: Apart from saving trips, passengers feel safer, if they do not ride alone with the driver;
- Strict controls of car occupation on HOV lanes;
- A high concentration of working places at central locations and a high concentration of commuters in residential areas;
- Long journey times;
- Convenient alternatives with the public transport;
- A bottleneck through which all commuters have to pass by.

He also found out that the informal carpooling system is based on a ratio of passengers to drivers of 2:1, while only 12 % alternately driver or take a lift (Oliphant 2008, p. 23).

3.3.21 Concept on "Less Cars in New York" (USA)

The New Zealand based company Trip Convergence Ltd. has developed a flexible commuter carpooling system using infrared or RFID (Radio Frequency Identification) for data capture and transmission. The system is illustrated within the concept study "Less cars in New York" (www.lesscarsinnewyork.com), but has not been tested in reality yet. The system is based on the analysis of informal carpooling practices in Washington and San Francisco, where the pull-factors are so strong that carpoolers have developed an informal carpooling system without any central organisation or service.

The basic idea of the concept is to design carpooling parking lots that both promote carpooling and control the usage. A special design allows each driver to flexibly decide on the spot, if he wants to give a lift to other passengers or to park his car and be a passenger (Fig. 3.2).

Drivers looking for passengers stay on the middle lane and pick-up potential passengers, who have parked their car in dedicated areas (e.g. “university”, “city centre”, airport etc.). In case of a passenger surplus, users who have already parked their car can drive onto the middle lane giving a lift.

Apart from the parking lot design, the system uses infrared or RFID technology for billing the trips and identifying users for security reasons. For this purpose, registered users get a personalised smart member card, which can be activated by a finger print at the beginning of each ride. Drivers also get an infrared car tag, which captures the signal identifying the passengers, and transmits the ID of the car and all passengers to a RFID reader at the car park exit. The reader sends this information to a database for checking the IDs and billing the passengers. Simultaneously, all occupants will see the names of the driver and the passengers in order to check if there have been any transmission problems. Figure 3.3 illustrates the system (according to www.lesscarsinnewyork.com):

The “Less Cars in New York” website proposes on registration costs of 25 $ as well as an annual fee of 150 $.These costs will, however, vary depending on sponsoring or public funding. Furthermore, the costs for the electronic membership card and the IR-tag account for 300 $, financed either by funds or an annual rental fee of 50 $. Besides, it is considered to charge a service fee of 1 $ and a parking fee of up to 5 $ per use. All in all these costs are considered as maximum costs, reflecting the actual expenses. All data on lifts identified by the RFID interface is documented on the user accounts in order to make the calculation of the costs transparent.

3.3.22 Nuride.com: Rewards for “Greener Trips” (USA)

Nuride.com is a bonus program aiming to avoid single car rides. With more than 60,000 members, it is one of the biggest reward programs supporting carpooling in the US. The principle of nuride is to avoid car rides and reward commuting and private trip alternatives as “greener trips” such as public transportation, carpool/ van pool, by foot or by bike and telecommuting. After signing up, registered users can earn nuride-points by reporting their “greener trips” and tracking their savings with regard to costs and CO_2-equivalents. These points can be redeemed for rewards like restaurant coupons, retailer discounts or tickets to shows and attractions. Moreover, nuride members can contact other members to ride or start a carpool. Nuride is for free, since it is sponsored by governments and companies that aim to support “greener trips” as well as businesses providing rewards. 56 % of the users are female and concentrate in Washington D.C. and New York City (with about a third of the users in each city).

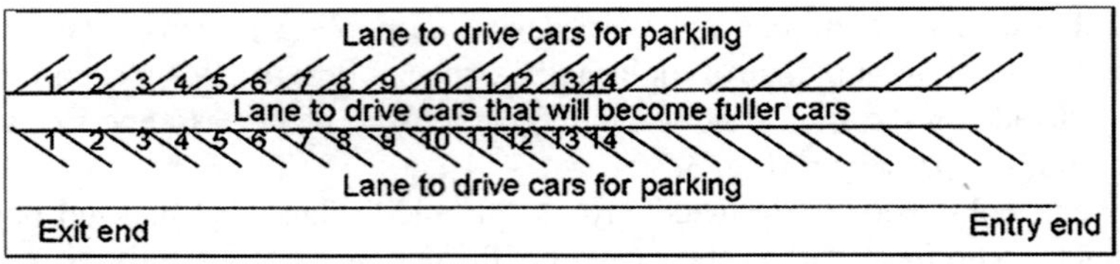

Fig. 3.2 Parking lot design for the "Less Cars in New York" concept

3.3.23 "HOV-Jockeys" in Jakarta (Indonesia)

Jakarta has a severe traffic problem: Each year the number of licensed cars and motor-bikes increases by 9.5 % and each day, traffic jams paralyse the transport on main streets during the rush hour. Due to the daily gridlock in Jakarta, so called "3-in-1" lanes were introduced in 2006, which only cars with more than three passengers are allowed to use during peak-hours. These lanes are either dedicated express lanes on motorways or whole streets in the city centre that are restricted to cars with at least three people.

As a response on the "3-in-1" lanes, residents of Jakarta have developed a particular kind of informal carpooling: So called "HOV-jockeys" wait at central junctions for single drivers paying them a small fee for getting into the car, which permits the driver to use "3-in-1" lanes and streets. For the mostly very poor HOV-jockeys this form of ridesharing is a possibility to earn a living (Onishi 2009). However, since HOV-jockeys undermine the intent of reducing traffic on the whole, they were prosecuted and charged by governmental institutions. In future, an electronic card system should support "legal" carpools and counteract "HOV-jockeying".

3.3.24 Megacarpool (Indien)

In order to fight the daily traffic jams, the government of Delhi has developed the "Megacarpool" concept (www.megacarpool.com) in cooperation with the Indian Institute of Technology in 2009. The Megacarpool system uses up-to-date technology for controlling and billing carpooling trips (Sharma and Sakhuja 2009). Thus, each Megacarpool member gets a GPS device and a MEGA CASH smart card, which is read by every passenger at the start and end of a ride. The route will be measured exactly by GPS and the data will be transmitted with GPRS to a database in order to calculate ridesharing credits. The rates are dependent on the number of passengers and span for drivers credits from 3.00 (one passenger) to 6.00 Rs/km (four passengers) resp. a passenger contribution of 3.50–2.00 Rs/km.

Apart from supporting the billing process, the GPS tracking serves enhances the safety of the trips and allows finding the car in case of a theft. The GPS device

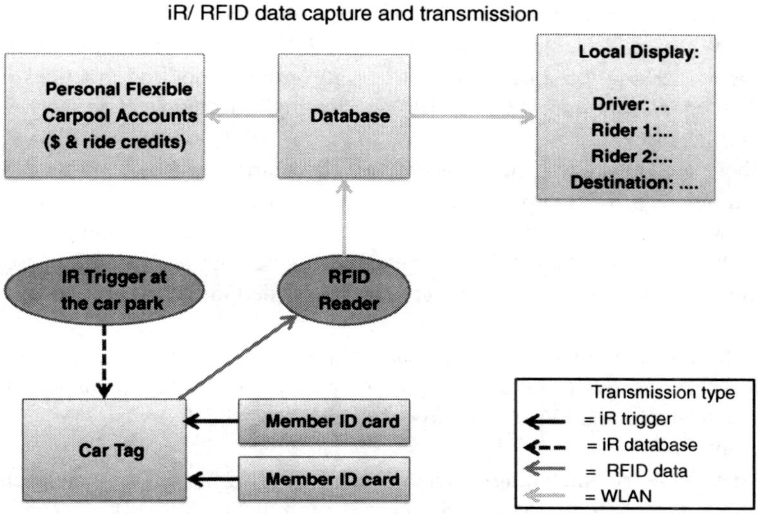

Fig. 3.3 Data transmission within the HOVER system for flexible carpooling

is also equipped with a "panic button", which can be used in case of accident or emergency. Applicants can register online by indicating name, email address, vehicle brand, mobile number and vehicle registration number. For getting the GPS device and MEGA CASH smart card, a Mega Carpool employee visits all applicants at home. The formal identification process includes checking the registration card, the driving license, the identity card as well as the citizenship. Member without a car can only participate as so called "ride matches", if a registered car owner recommends them.

The actual ridematching service is yet only available for fixed routes that are either indicated during the registration process or transferred by SMS, internet or a call at the call centre. In case of a matching partner, the Megacarpool call centre informs both members and exchanges the contact data. Megacarpool also plans to offer an ad-hoc matching in the future. The Megacarpool website does not give any information on the number registrations or lifts.

3.4 Conclusion on the Conception of Carpooling Offers

Comparing international carpool practices and services, it is obvious that national and local framework conditions are the deciding factors for the demand for ridesharing. The strongest pull factors are the existence of HOV lanes (HOV = High Occupancy Vehicles), which only cars with more than two or three passengers are allowed to use. Further pull factors are a lack of appropriate public transportation options as well as a high traffic volume on certain routes or bottlenecks like tolled bridges. Where pull factors accumulate, e.g. HOV lanes, strong congestions and a shortage of parking lots in central districts, informal carpooling practices like the slug-lines in Washington D.C. or "HOV-jockeying" in Jakarta even work without any external support.

If there are only few general incentives, ridesharing can be supported by adequate strategies providing push factors for carpooling. With regards to increasingly flexible working times, carpooling schemes have to meet the challenge of providing flexible and individual mobility that can compete with individual car use.

In this context, ridesharing practices can be divided into two categories:

- On-site ridematching on dedicated carpooling sites (e.g. slug-lines in Washington D.C., HOV-jockeys in Jakarta, carpooling parking lots in Saarland/Germany and Ottawa/Canada, the CARLOS carpooling call box, the "Less cars in New York" concept) and
- Internet-based ridematching services (M21 Fahrplus, frauenfahrgemeinschaft.de, mitfahrgelegenheit.de, Move Bremen, Commuters' Network Saxony, City-Netz, Easy Rider, Covoiturage.fr, Liftshare.com, DriJo, CommuteChoice Ride Sharing, GoLoco, Jayride, Zimride, Nuride.com, Megacarpool.com).

Depending on demand and offer, on site ridematching always offers the opportunity of spontaneous and flexible ridesharing. Offers that are combined with parking lots (slug-lines in Washington D.C., carpooling parking lots in Saarland/ Germany and Ottawa/Canada, the "Less cars in New York" concept) moreover offer the choice of parking the own car and be a passenger in a carpool. These practices obviously avoid a single car use and save costs as well as CO_2 emissions.

In contrast, passengers matched by internet-based carpooling services are not necessarily drivers, but potentially also use public transport. Here, saving costs and a higher comfort in comparison to public transport are the main motivation for carpooling, but these trips do not result in CO_2 emission savings. Still, some passengers using internet-based services are definitely also car owners, who refrain from taking their own car.

The most obvious problem of internet-based services is, though, that they are not flexible enough for the addressed target group. Internet-based services always require determining routes, travel times and roles (driver or passenger) in advance. Phone based carpooling services (M21 FahrPlus, Easy Rider, Move Bremen) provide much more flexibility, since matching partners get personally into contact and can negotiate time and pick-up points. However, phone based services are still dependent on a preliminary matching procedure as precondition for connecting potential carpooling partners. Actually, finding an adequate pre-selection procedure has been one of the main obstacles for the presented phone based carpooling services, which did not succeed in surviving the pilot project phase.

Today, the spread of smartphones and mobile internet promises to bring flexible carpooling to a new dimension, since mobile applications as e.g. offered by Comuto.com or Avego.com offer a simple still internet-based solution for real-time matching and for contacting matching partners. In future, GPS based systems such as e.g. used by Megacarpool will open up further, more user-adapted options for flexible carpooling services.

Nevertheless, the technology itself is not a sufficient push-factor for motivating commuters to carpool, if there are no additional pull-factors like HOV-lanes or an inadequate connection to the public transportation network. Therefore, many services offer further incentives like special parking lots for carpooler (Harvard CommuterChoice Program), bonus programs (nuride, Megacarpool.com), the connection to social networks like Twitter or Facebook (Zimride, GoLoCo, Megacarpool.com). Especially the connection to virtual social networks is crucial for carpooling services, since they enhance trust between the carpooling partners, which is an essential incentive for sharing a ride in a private car.

There are considerable cultural differences regarding financial contributions for the drivers. For instance, in the USA and Canada, it is unusual to share costs (e.g. OttawaRideMatch.com warns of "illegal bandits" demanding a contribution for the driver). In contrast, sharing driving costs is an important argument for carpoolers in Europe. Yet, European carpoolers prefer paying "behind the scene", e.g. via a member account or like within the CARLOS project via a ticketing system. As the Easy Rider project has shown, an adequate billing of the trips is, however, critical for passengers.

Further projects have shown that concentrating marketing activities passengers is another important success factor. According to Oliphant (2008) self-organised slug-lines with have a ratio of 12 % "pure" drivers to "pure" 60 % and 28 % "alternating user", who sometimes drive and sometimes take a lift. This ratio, thus, seems to represent a good balance between supply and demand.

With respect to internet based services, service providers should not have great expectations regarding daily matchings. Although, for instance, Jayride is acknowledged as a successful carpooling service, only 10–15 % of all about 6000 registered members have ever been matched and most of them only one or two times.

The presented projects had different experiences with marketing activities.

The CARLOS evaluators concluded that the main target group (teenagers in rural areas) were not sufficiently informed on the service, which has been reflected by only a small number of users. The carpooling project at the University of Oldenburg revealed that even a variety of marketing activities, a high level of awareness and an explicit encouragement of the target group did not result in actual carpooling (Sonnenschein et al. 2005). However, an appropriate marketing strategy is nevertheless a central precondition for implementing a new carpooling service.

In summary, spatial frameworks are the strongest pull factors to carpool. Thus, carpooling lanes and parking lots are important incentives for carpooling services. The Move-project has moreover shown that particularly corporate carpooling schemes including an appropriate parking management and targeted cooperation can be an important incentive for carpooling, particularly when accompanied by a detailed analysis of local mobility patterns or carsharing options. This, however, requires a complex and comprehensive strategy that should moreover be dynamically adapted to the carpoolers' needs and changing frameworks. Therefore, service providers have to find an adequate balance between costs and benefits in order to establish a successful, but feasible carpooling concept.

Chapter 4
The CARRIVA Ridesharing Club

The CARRIVA Ridesharing Club (CARRIVA-Mitfahrclub) was founded by the support of fundings from the Federal Ministry of Transport, Building and Urban Development in 2007. The carpooling service predominantly addresses about 70,000 commuters, who work at the Frankfurt airport. The first step was to establish an internet-based service that should be expanded by a phone based ridematching. The main goal of the accompanying empirical research was to analyse the user acceptance and motivations regarding an internet or mobile phone based carpooling concept. Furthermore, the scientific team has evaluated the implementation process in order to adapt marketing strategies to the users' preferences.

The following sections deal with the local mobility structures at the Frankfurt Airport (Sect. 4.1), the description of the CARRIVA service concept (Sect. 4.2), and finally the concept of the scientific evaluation (Sect. 4.3).

4.1 Analysis of Transport and Mobility Structures at the Frankfurt Airport

For evaluating the implementation phase of the CARRIVA Ridesharing Club, it is firstly important to analyse the socio-economic and geographical framework at the Frankfurt airport area, which are on the one hand determined by the traffic infrastructure around the Frankfurt airport and on the other hand by the mobility patterns of the commuters. On this basis, we are able to extract the general potentials of commuter carpool services at the Frankfurt airport.

V. Handke and H. Jonuschat, *Flexible Ridesharing*,

DOI: 10.1007/978-3-642-11345-1_4,

Table 4.1 Car stock and annual kilometers travelled by public transport in the administrative district of Darmstadt for the 1st quarter 2007

District	Number of residents	Number of cars per 1,000 residents	km travelled by public transport (in 1,000 km)	km travelled by public transport/ per capita	Type
Bergstraße	264.622	570	669	13,87	A
Main-Taunus district	224.713	704	164	22,98	A
Odenwald district	99.28	555	725	27,45	A
Offenbach	336.518	556	720	19,97	A
Rheingau-Taunus district	184.077	580	563	8,49	A
Groß-Gerau	252.421	560	446	29,50	B
Main-Kinzig district	408.546	538	394	30,34	B
Wetterau district	299.135	544	371	41,36	B
Darmstadt-Dieburg	289.087	554	954	41,35	B
Hochtaunus district	226.453	669	58	35,58	B
Darmstadt city	141.471	468	184	29,57	C
Frankfurt/Main city	654.172	444	243	52,35	C
Offenbach/Main city	117.923	415	973	50,65	C
Wiesbaden city	274.771	486	116	33,18	C
Average	*269.513*	*545*	*(6524 o. FFM)*	*31,19*	

4.1.1 Transport Structure within the Area Covered by CARRIVA

As the previous chapter have shown, potentials for avoiding traffic by carpooling schemes depends predominantly on the group of frequent car drivers and potential public transport users, which represent approximately half of the "mobile" population over 14 years (see Fig. 2.5). Both groups differ according to regional context factors, e.g. the public transport network or main commuter routes. Here, the decisive parameter is the number of cars in private households. In the administrative district of Darmstadt, which includes the Frankfurt airport, this value varies from 444 cars in Frankfurt city up to 770 cars per 1,000 residents in the Main-Taunus region (Statistisches Landesamt 2009).

Number of cars in private households indirectly is closely linked to the quality of public transportation network, which is obviously much denser in Frankfurt city than in the Main-Taunus region. Table 4.1 presents the main data related to car traffic (Statistisches Landesamt 2008):

Although this data is only a rough indicator on the quality of the local public transport network, it is obvious that in some districts the average number of private cars per capita is high, while the travelled distance by public transport is low (Type A). Here, carpooling is the main alternative to commuting with the own car. On the contrary, in cities like Frankfurt, Offenbach or Wiesbaden, residents have

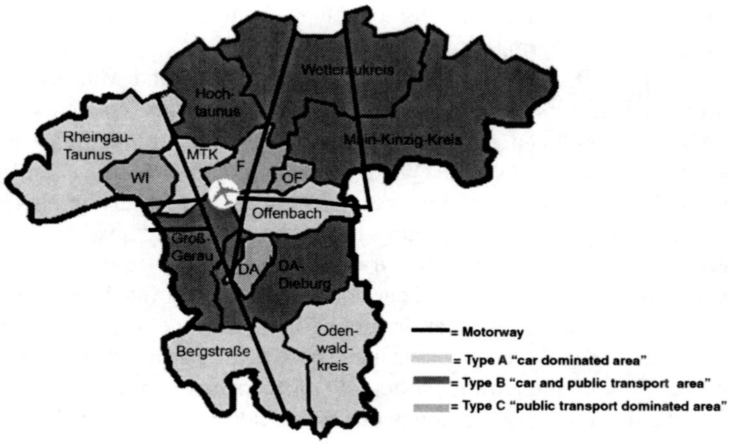

Fig. 4.1 Mobility types in the CARRIVA area

less cars than the average and the number of kilometres travelled by public transport is relatively high (Type B). In these areas, public transport seems to be an adequate alternative to individual car use. Still, residents of Type B districts could have nevertheless an interest in carpooling, since public transport can be used e.g. in case of problems in finding a carpooling partner or for the ride back home.

Furthermore, there are regions with a number of cars per residents as well as a good public transport network (Type C). Particularly in these regions, drivers and passengers can change flexibly between car use, public transportation and carpooling.

Figure 4.1 illustrates that the regions adjacent to the Frankfurt airport are quite diverse with regards to the number of cars per capita and the density of the public transport network. As a consequence, promotion strategies for CARRIVA should consider the dominating mobility type (car/public transport dominated or both) per region in order to be effective. Particularly Type A (= car dominated) and Type B (= good car and public transportation network) districts are of interest, while urban districts of Frankfurt, Wiesbaden, Darmstadt and Offenburg city are less attractive for ridesharing, since public transport is a good alternative to car use.

Due to their vicinity to the airport and a good access to the motorway, it seems to be most effective to encourage drivers living in the districts Groß-Gerau, Offenbach and Main-Taunus to join the CARRIVA Ridesharing Club, since other commuters will pass by these districts on their way to the airport and residents do not intensively use the public transport (less than 30 km quarterly or about 10 km per month). Here, it is most probable that employees will abstain from taking their car and take a lift instead. Moreover, living along the main routes to the airport their chances to get a lift will be comparatively high. In these districts, the main target group are "ridesharing-passengers".

The three Type B districts Hochtaunus, Wetteraukreis and Main-Kinzig-Kreis are also of particular interest for a ridesharing service, since residents have both good public transport and motorway access. Therefore, ridesharers supposably have the opportunity to use CARRIVA only one-way and travel the other by public transport. This also enhances the chance of using ridesharing as an alternative to single car use. The main target group in these districts are "flexible ridesharers" that alternately use ridesharing, public transport and their own car.

Finally, in remote, less dense and car dominated districts like Rheingau-Taunus, Odenwaldkreis and Bergstraße, car drivers will only change to ridesharing, if they have a guaranteed ride home or a constant ridesharing partner. However, they are still interesting for giving a lift to carpooling partners on the route to the airport. In these districts, the main target group is therefore "ridesharing drivers".

Estimating a general potential for ridesharing on basis of district data relating to the number of car owners and the density of the public transport system is only a very rough method to determine target areas for a promotion strategy. In fact, individual preferences and circumstances can of course result in a high motivation to use ridesharing in urban, public transport dominated areas. On the contrary, people living near a train station in otherwise car dominated areas will also rather prefer public transport to ridesharing. Nevertheless, a thorough analysis on the transport options can be a useful starting point for targeted marketing strategies, but also for infrastructure measures such as establishing "ridesharing pick-up points" or carpooling parking lots.

A more detailed analysis of route convergences resulting in a good matching probability needs information on the point of departure, i.e. the private address. If commuters to the Frankfurt airport live nearby, chances for ridesharing are obviously the best. However, drivers can also "collect" passengers on their route, if this does not result in a longer detour. Finally, the Frankfurt Airport covers a wide area and bus trips from Terminal 1 to the Cargo City South can also take about 45 min. Therefore, the final destination at the airport is also important information for evaluating matching chances.

Summarizing, in order to maximize matching chances, the drivers' and passengers' willingness to accept a detour—both at the departure point and at the destination—is a central factor for their willingness to use a ridesharing service. As explained in Sect. 3.2, not only the route convergence, but also the time convergence is a key factor for finding matching ridesharing partners.

4.1.2 Commuting Times of Potential CARRIVA Users

Rush hours characterized by a high traffic density on certain routes indicate a good potential for ridesharing options. In times of an increasing flexibility of work schemes, rush hours are harder to predict, but they are still visible in terms of regular traffic jams. In general, on can determine three general working schemes that have an influence on the commuting times:

1. Office working times that are often arranged as "flextime" schemes, allowing flexible working times around a certain daily attendance "core period" (e.g. between 11 a.m. and 3 p.m.)
2. Shift working with fixed working times that periodically change
3. Completely flexible working times (also regarding the working days) that are particularly relevant for part time employees, remote workers or associated employees, but also for field staff or managers travelling a lot.

Either at Lufthansa (35,000 employees at it's Frankfurt base in total) or at Fraport (18,000 employees in total), two-third of the staff are employed as shift workers and one-third as "flexworkers". From this, we can conclude that a third of all employees at the Frankfurt airport could find a fix ridesharing partner, since only "flexworkers" are able to determine fixed daily commuting times. These flexworkers are also an interesting target group for ridesharing services, since they can adapt their travelling times to a certain extent to the travelling times of a potential ridesharing partner. This means, however, that in this case, commuters have to abandon their option of flexible working times.

Two-thirds of the airport staff, i.e. about 47,000 employees, can only determine their commuting times, when they get their quarterly shift plan. These people cannot determine fix travel times in general, but they at least know some months or weeks in advance the exact time of arrival at the airport. This entails that shift-workers using an internet-based ridesharing service have to regularly update their commuting patterns.

For an internet-based ridesharing service, it is a big challenge to address flexible working times by allowing users to indicate travelling periods instead of fix travel times, since the related database needs to match time ranges instead of a given time. This requires a much more sophisticated algorithm, which is why existing internet-based ridesharing services did not integrate the option to indicate flexible commuting times yet. Consequently, all of the 14,000 employees working as flexworkers at Frankfurt airport can only use a common internet-based ridesharing services, when indicating fixed travel times in advance—although they are in fact much more flexible. Therefore, with respect to the growing number of flexible workers, there is an urgent need for a dynamic ridesharing service reflecting daily changing commuting times on the one hand.

On the other hand, only a dynamic ridesharing service can use the ridesharing potentials to full capacity, since time ranges maximize the matching probability considerably: A commuter travelling at 9 a.m. will not be matched with someone travelling at 9.45 a.m., but ridesharing partners travelling between 9 a.m. and 10 a.m. and will be connected. But even if a dynamic ridesharing service offers the option to determine time ranges instead of a fix travel time, the actual travel time is uncertain. If a flextime worker, for example, indicates to usually depart between 8 a.m. and 10 a.m., and he starts at 8.15 a.m., he will not be able to give a lift at 8.30 a.m. anymore.

In summary, both flextime workers and employees with completely flexible working times need a dynamic ridesharing service in order to profit from the time autonomy given by their working scheme. For shift workers a dynamic ridesharing service is also more convenient than a standard internet-based service, since they can spontaneously decide to give or taking a lift. For developing a dynamic ridesharing service, it is therefore essential to determine the actual commuting times and tolerance in order to maximize the matching probability that determines the success of a ridesharing service.

4.1.3 Potentials for a Dynamic Ridesharing Service at the Frankfurt Airport

As stated before, the matching success is dependent on a convergence both of the routes and times of travelling ridesharing partners. The ideal situation is that the driver and passenger live nearby and work at the same times. However, most employees work flexibly, and at least drivers can make detours or pick-up a passenger on the route. Thus, the success of a ridesharing service is closely related to the willingness of all partners to adapt their travelling times to ridesharing options as well as to the drivers' willingness to pick-up passengers on or near the route.

Apart from a time and route convergence, the third main determinant for a ride-matching success is a good balance of drivers and passengers, since two matching drivers or two matching passengers sharing the same travelling time and route still cannot form a ridesharing community. Therefore, a dynamic ridesharing systems has to reflect the ridesharing partners' status: Either exclusively driver or passenger or "flexible partners", who are willing to drive, but also to be a passenger. Since one driver can give a lift to up to three passengers, the ratio between drivers and (potential) passengers should be about 1:3. This fact is often neglected by many ridesharing services that put a lot of efforts in recruiting drivers instead and eventually end up with a passenger shortage.

As explained in the previous chapters, Groß-Gerau, Offenbach and Main-Taunus seem to be the most promising districts for promotion activities. While shift workers (two-thirds of all employees) can use an internet-based ridesharing service, flexi-time workers (one-third of all employees) are more interested in flexible, dynamic service concepts. In comparison to many non-European regions, there is no strong demand for ridesharing services in Germany, since in most parts of the country, the public transport network is dense and parking or congestions problems are tolerable. Furthermore, there are no strong (financial) incentives for ridesharing like HOV lanes or toll roads or bridges. However, in a national or European comparison, the situation at the Frankfurt airport is comparably good for ridesharing services:

- The airport is a central destination. Apart from the Cargo City South, where about 6,000 people work (Gauler/Gretz 1999, p. 6), most parts of the airport

(Terminal 1, Terminal 2 etc.) are in the same area and easily to reach by car or by public transport.

- A dense motorway network links the airport to the surrounding regions. In some of the adjacent districts the car is the main transport means, since the public transport offer is not optimal.
- Most districts, however, have at least a basic connection to the public transport (e.g. one bus per hour), which guarantees an alternative to car use.
- The parking situation at the airport and congestions on streets and motorways around the airport are strong pulling factors for carpooling. There are in fact enough parking lots for all employees, but many parking sites are located in a big distance to the main working sites and pedestrians have to pass unattractive grounds.

In Germany, costs for fuels, but also for public transport are moreover relatively high, and carpooling could therefore be an attractive means to reduce mobility costs. However, some employers such as Lufthansa in Frankfurt offer free "jobtickets" for the public transport to their employees, and this considerably undermines the financial advantage of ridesharing.

One must take into account that in spite of a multitude of ridesharing funding programs and new ridesharing services during the last decade, many services are based on "dead databases" containing old, outdated offers of drivers who have never or only few times found a passenger. This is also due to the fact that flexible ridesharing concepts are still an exception, and none of these concepts has survived the pilot project phase up to date. Though, with regards to an increasing flexibility of working schemes, and the spread of new communication technologies (mobile internet, GPS etc.), the situation for launching a flexible ridesharing service like CARRIVA is extraordinary good.

4.2 The Concept of the CARRIVA Ridesharing Club

CARRIVA was explicitly designed as dynamic, flexible ridesharing service. As an registered association, CARRIVA is a non-profit organization. The concept is based on the idea of a club in order to emphasize the community character and to encourage social and informal relationships between the ridesharing partners. Thus, CARRIVA users are in fact club members. They do not have to pay a membership fee during the launching phase, but at a later stage, small membership fees can help to support service organization on a long term.

The unique selling factor of CARRIVA has been and still is to use the mobile phone as communication tool for matching ridesharing partners on the spot. Using a mobile phone drivers or passengers can look for ridesharing partners directly before departing, which indeed allows flexible and spontaneous ridesharing. In addition, CARRIVA also offers an internet-based ridesharing service to customer groups who like to or have to plan their trips in advance. Members with regular travelling times can set their data in a personal "ride calendar" in the private section of the CARRIVA homepage.

In order to avoid frustration, flexible matching options, e.g. by mobile phone need a critical mass of users at any time, since users expect an immediate information on an adequate match in case of a personal call. In contrast, users show a higher tolerance for time-shifted or even no information at all when using internet-based communication. This is due to the fact that internet communication is per se time- shifted, and users do not demand a direct reply. When calling by phone, we however indeed expect a direct answer and information.

As a consequence, a spontaneous ridematching by phone needs a sufficient pool of riding opportunities at any time, while an internet-based inquiry can be processed with a time lag and can moreover be daily repeated without any efforts. Therefore, a spontaneous ridematching service needs a critical number of users[1] to guarantee an adequate matching success directly by phone. On the opposite, an internet-based service could also work with only few members, who send a single, but daily updated inquiry and only get information, when a matching partner has been found. Therefore, an internet-based service represents a good basis to get accustomed with a dynamic ridesharing service during the launching phase with only few members. After having reached a critical mass, the service's database can be used as "backbone" of constant offers, which could be matched with spontaneous calls in regions with a sufficient concentration of offers. In case of a broad offer of constant and spontaneous ride opportunities, even two spontaneous calls can be matched. For both spontaneous and regular users, a higher number of ride opportunities result in a better chance to get a lift or a passenger, but do not change the communication process by phone or internet.

In summary, the CARRIVA offer combining both a spontaneous phone-based and a regular internet-based ridematching, will change from a high share of regular offers to an increasing number of spontaneous offers over time, and marketing strategies as well as operation processes have to dynamically adapt to each stage of the service. In the following sections, we explain the main operation processes as available during the launching phase (state of 2010).

4.2.1 Registration Procedure of CARRIVA

New members have to register on the CARRIVA homepage www.carriva.org by completing their personal profile. Here, applicants can indicate their personal data, their preferences for the matching process and their starting point and destination.

[1] The CARRIVA initiator expects a sufficient matching success at a critical mass of more than 4,000 members.

For a registration, it is required to leave a mobile phone number, which is used as user name, and to choose a password for the login. On the registration page, interested applicants can also read the terms and conditions of CARRIVA.

4.2.1.1 Personal Data

In addition to the mandatory mobile phone number, users can declare their e-mail-address, first and family name, birthday, postal address and home phone number. Moreover, members can leave their bank account data, which is required for an automated billing process that is planned for the future.

4.2.1.2 Ridematching Preferences

With regards to preferences for the ridematching process, members can indicate, if they intend to give or to take a lift. Once defined as driver or passenger, users can determine their preferences in terms of sex or smoking, i.e. if they prefer women or men as passengers or drivers, if they like to smoke during the ride or if they prefer a non-smoking car. Furthermore, users can indicate if they accept a variation of the route and how many kilometers are acceptable. As a preparation for the spontaneous matching by mobile phone, members can indicate, how much time in advance they will probably call CARRIVA.

4.2.1.3 Routes

The personal profile contains a part called "my addresses", where users can define seven different addresses as starting points or destinations. All addresses get a specified number that is used both by the spontaneous and the internet-based database system. The first address (No. 1) is determined as the airport address that however is specified by a set of 29 locations to choose from. The database consequently calculates each route to work e.g. as trip from address 2 to address 1. Still, users can also define a specific route on a map displayed on the page "my route". Finally, users can define one address as destination for a potential future guaranteed ride home offer.

4.2.2 Internet-Based Ridematching by CARRIVA

After registration, users can directly use the internet-based ridematching service. During the launching phase, the ridematching process starts, when a user states his time preferences. These preferences are patched with all offers in terms of route, preferences and times. In case of a successful match, both partners get a message with the contact data of the other matching partner.

The matching of the route does not only use the addresses indicated during the registration process, but also a simulation that allows matching addresses on the route as well. The simulation also takes passengers' addresses into account with a distance of 500 m to the route. If two members share the same route, they will however not necessarily be a good match, since also the driving times have to coincide. CARRIVA uses a "trip calendar" for matching trips by time on the one hand and a "personal schedule" on the other hand.

4.2.2.1 Using the Trip Calendar

If a registered member has fixed commuting times, he can use the "trip calendar" for publishing lift offers or requests. The trip calendar functions as "backbone" of the service as a fix database on regular offers. It offers three different ways to fill in personal commuting times:

1. *Shift plan*: The shift plan is based on shift patterns that have to be personally defined by the user e.g. A = early shift from 6 a.m. to 3 p.m. both as driver, B = day shift, 10 a.m. as driver to the airport, and at 7.p.m. back as passenger etc.). These shift patterns can be dragged and dropped on a calendar, marking the date with the user's shift.
2. *Regular trips*: CARRIVA members can announce "regular trips"on certain days. After having published the first "regular trip", a pop-up-window offers the possibility to transfer the dates for each day. Though, it is also possible to indicate different travel times for each weekday.
3. *Current trips*: The trip calendar contains a part, where users can announce "current trips"on specific dates. This is particular relevant for people who irregularly make a trip to the airport.

In addition, each user has to indicate his status as driver or passenger. If a person is exclusively driver or passenger, all announcements are marked according to the given category. Users however have still the opportunity to mark trips as both, driver and passenger. This application addresses mainly for those members who usually drive with the car, but are also willing to carpool as a passenger. If one of these members has taken a lift on the way to work, the database deletes the opportunity to give a lift for the way back home. If he has given a lift to a passenger or has not been matched at all, the system assumes that he is driven by car and will not be a passenger on the return trip.

After having completed the trip calendar, the system instantly matches the data with the database. This results in two possible outcomes:

1. There are suitable drivers or passengers for the given route and time. In this case, a pop-up-windows suggests contacting one of the relevant ridesharing partners by CARRIVA.
2. There are no matching partners at all, which is announced by a pop-up window.

All trips listed in the trip calendar are the basis for the "personal schedule" that represents an alternative particularly for members with irregular commuting patterns to find a matching partner.

4.2.2.2 The Personal Schedule

Each user can create a "personal schedule", which contains the travel times of all trips on the same route for six days in advance. Here, one can not only find travel times of all members who live nearby, but also those of "potential passengers" who live on the route, and of all "potential drivers" who pass by. Users can contact each potential driver or passenger separately via CARRIVA or send a message to all members within the personal schedule. Here, he can choose, if he wants to publish his e-mail-address and/or telephone number.

Similar to a bus schedule, the personal schedule lists potential travel times on the commuting route as listed in the trip calendar. By choosing a regular trip from the personal schedule up to 6 days in advance, users can contact a regular, trip calendar driver or passenger—even if they did not fill in the trip calendar themselves. This allows that according to the schedule at least the contacting person can flexibly choose the departure time.

Apart from that, members can use a link to send a message to all drivers and passengers on the route, regardless if they have published their travel times by the trip calendar or not. Within this message, members can inform all recipients on their travel times and their contact data. This offers the opportunity to find a carpooling partner, if a person cannot or does not want to determine any trips in advance.

Summarizing, in contrast to the trip calendar, the personal schedule potentially offers the opportunity for flexible ridesharing, which is however dependent on the communication processes between the ridesharing partners. If the driver and the passenger communicate via internet, the lead time for arranging the trip is much longer than in case of a phone contact. If a driver or passenger has published a mobile phone number or uses a smartphone, the personal schedule theoretically allows contacts only few minutes in advance. Thus, both the personal schedule already offers the option of spontaneous, ad-hoc ridematching, at least if both carpooling partners use the mobile phone for communication. The CARRIVA concept however goes beyond this point with an explicit concept for ridematching by phone.

4.2.3 Ridematching by Mobile Phone

The unique selling point of CARRIVA is indeed the option to spontaneously find a ridesharing partner by the use of the mobile phone. Therefore, the whole service concept is adapted to the objective to match carpooling partners dynamically by

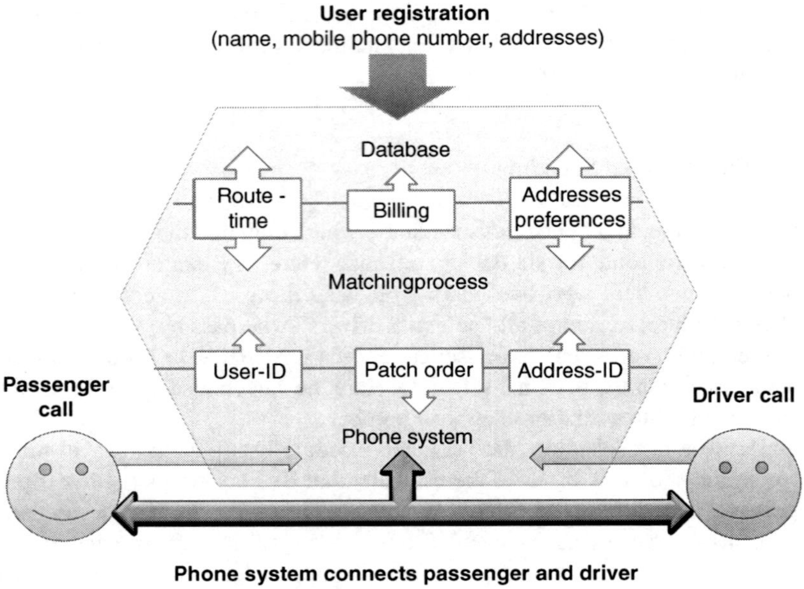

Fig. 4.2 Mobile phone based matching process of CARRIVA

phone. For instance, user names are synonymous with the mobile phone numbers, which allows identifying each member directly by the phone number. The phone based matching system also uses the personal address numbers defined during the registration process.

For using the phone based system, members have to call CARRIVA either as driver or as passenger (069 99999 101 or −100), adding an address number combination defining the route. Since 1 is always defined as airport destination and 2 usually as home address, a combination of −21 indicates a trip to the airport and −12 back home.

If a passenger needs a lift and a matching driver is available, the passenger is directly connected to the driver's mobile phone. If a driver on the contrary offers a lift and a matching passenger is available, the passenger gets a short message to contact CARRIVA in order to be connected to the driver. Consequently, in both cases, the passenger has to pay for the call. Figure 4.2 illustrates the phone based matching process:

If the user has not changed his personal preferences, each abbreviated dialing is calculated as a trip within the next 20 min, saved in the same database as the trip calendar data. As a consequence, the phone based system contacts both members that have also spontaneously called some minutes ago or members who have published an inquiry a long time before. Thus, the phone based matching system and the internet based matching systems are entwined. However, CARRIVA can only connect matching partners, if one of them has spontaneous called CARRIVA.

4.2.4 Strengths and Weaknesses of the CARRIVA Concept

CARRIVA distinguishes from other ridesharing services by the fact that it principally offers a spontaneous ridematching by using a mobile phone that is a widespread and ubiquitous communication medium. Only carpooling parking lots or pick-up-points or services based on mobile internet resp. smartphones can guarantee a similar realtime ridematching. At the moment, regular mobile phones are still much more spread than mobile internet compatible smartphones and the mobile reception is much better as well. Equally, a mobile phone call is much more comfortable and flexible than meeting physically at a carpooling pick-up point.

Since a phone based ridematching allows a definite identification, it can furthermore be connected to both automated billing and monitoring processes. In this respect, the anonymity of internet based ridesharing services is a great deficit, since feedbacks on matches, actual lifts etc. and a detailed analysis of carpooling activities is not possible. An automated billing process, shifting credits form one member account to another according to the route, cannot be organized by internet neither. However, this is indeed a prominent incentive to use a carpool service in comparison to privately organized carpools. Summarizing, CARRIVA's phone based ridematching service offers two important strengths in comparison to other services:

- *An automated billing process*: Credits shifted from passenger to driver are billed "in the background" on basis of the route that has been matched. Due to this, drivers do not have to collect (preferably the exact) cash money from the passengers. This makes ridesharing a lot more comfortable for the users, but also allows levying a certain share of the contribution as fee for the service activities in order to finance operating costs. Here, it is important to distinguish between a successful match and the actual ride.
- *Monitoring ridesharing activities*: The personalized phone based matching process allows analyzing all ridesharing activities in detail. As a consequence, the service provider can dynamically adapt the service to the needs and mobility patterns of the users and develop targeted marketing strategies for e.g. certain user groups or regions.

As explained before, a phone based ridematching however should only be introduced when the service has reached a certain critical mass of registered members in order to avoid frustrating inquiries. Therefore, an internet-based ridematching service is a perfect intermediate offer to acquire members during the launching phase. It is important to accept that launching the service is a dynamic process that starts with an internet based ridematching, but can be gradually expanded by a phone based ridematching according to regional concentrations of users, additional service options etc. A consecutive introduction of service elements furthermore allows adapting and improving the service gradually to user demands and specific conditions at each location.

However, one must take into account that internet-based ridematching services compete with several different local, regional, national or international carpooling services. It is therefore important to communicate a unique service option to the users. Moreover, matching two ridesharing partners with a similar profile always bears the risk that these partners form a fix carpooling group, and do not need a ridematching service anymore. Yet, fix carpooling communities can still be service customers, since they use the service for further trips or for acquiring further passengers.

The main weakness of the CARRIVA phone-based service concept is that the ridematching communication process favors drivers over passengers who have to pay the phone costs, although carpooling services need a driver-passenger ratio of 1:2 or more. Furthermore, each phone call is interpreted as a specific instant of time, 10 min after the call. This is due to the fact that the database calculates on basis of periods of 20 min, calculating a tolerance of 10 min before and after a specific, assumed departure time indicated either online or by a phone call. Whereas departure times listed in the trip calendar will get a message on a successful match some time in advance, a phone call inquiry can only be connected during the next 20 min to other potential carpooling partners. It is a great technical challenge to integrate inquiries by phone of both people who would like to give or get a lift within the next 20 min and people who are flexible during the next hour. This also emphasizes the need for a flexible and dynamic ridematching concept that connects not only spontaneously calling partners, but also instant calls with trip calendar entries that were perhaps made some weeks ago.

In total, most of the weaknesses of the CARRIVA concept are related to technical restrictions or systematic challenges that not only apply to the CARRIVA concept, but to all internet- or phone-based ridematching services. Thus, the evaluation of the launching phase aims to find out to which degree users tolerate service related restrictions, and to analyze the main ridematching characteristics that determine the acceptance of ridematching services and the motivation to use them.

4.3 Evaluation Concept for the CARRIVA Launching Phase

All initiatives that aim to support carpooling amongst commuters face the challenge to offer a high flexibility, but also a certain security of finding a successful match. Fortunately, both preconditions are closely connected: Flexible departure times theoretically involve a higher number of potential matching inquiries. In practice, however, a flexible mobility pattern entails a complex ridematching system based on a dynamic database that monitors actual riding times continuously. Here, three user types have to be taken into account:

1. "Permanent users" that have more or less fix commuting times that can be planned in advance,

2. "Partial flexible users" that can determine a certain time range for departure times such as flexi-time employees
3. "Completely flexible users" that like to spontaneously decide their departure time as well as if they want to give or take a lift at all.

The CARRIVA system aims to explicitly establish an attractive ridesharing service for both flexible user groups, since existing internet based services do not respond to their demands yet. Indeed, a lack of flexibility is the main reason why, in spite of a high number of registered members, most internet-based services manage to actually initiate only few shared rides. It is therefore obvious that there is indeed an urgent need for a real dynamic ridesharing system in order to support the idea of carpooling in general.

For a ridesharing service, "permanent users" with fix daily commuting times represent a high planning reliability. If they, however, find a matching "permanent user" as ridesharing partner, both are not "matchable" anymore, unless they can offer a lift for further passengers. In contrast, flexible users are technically an important customer group, since they have to constantly use the service offer in order to get or to offer a lift. Thus, flexible users spontaneously determining their departure times as well as their role as driver or passenger are the main target group of CARRIVA's dynamic ridesharing service. They are as well the main target group for climate protection strategies, since drivers who are willing to leave their car and take a lift, but do not want to give up the flexibility of individual car transport belong to this user group. Consequently, the evaluation of the CARRIVA launching phase focuses on partial and completely flexible users, their characteristics and their expectancies and demands for a flexible ridesharing system. Apart from that, the evaluation aims to give advice on constantly adapting the ridematching system to the needs and mobility patterns of all users.

Currently, most carpooling services are internet-based and address exclusively commuters with fix or at least projectable travel times. These user groups however only represent about a third of all potential users. Consequently, ridesharing services have to adapt their service offer to flexible users e.g. by a phone based matching in order to tap the full customer potential. Figure 4.3 illustrates this argument:

Since an internet based service is technically less complex and less demanding than a phone based service, it is recommended to plan a progressive expansion of the phone based service. If permanent lift offers and inquiries accumulate in a region, it is comparably "safe" to launch the phone based service by communicating to all users within these regions that they can also call CARRIVA for finding a carpooling partner. At this stage, spontaneously calling "flexible users" are predominantly matched with "permanent users" using the trip calendar. However, even at this stage, it is possible that spontaneous calls are matched with other spontaneous calls. Consequently, at later stages, only the probability of getting a match increases, regardless if one of the matching partners has made a spontaneous call or set his preference in advance by internet.

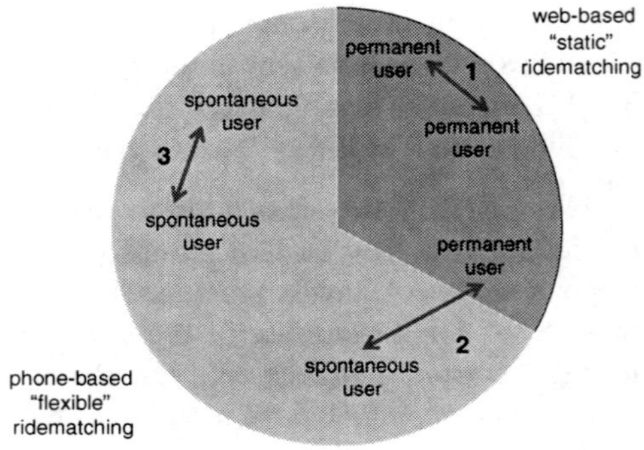

Fig. 4.3 Matching chances between user types

Thus, the first launching stage of launching the CARRIVA offer is exclusively internet-based, while the second test phase integrates a phone-based matching in several regions with a broad offer and the third expansion stage aims at expanding the offer to all users. The evaluation concept consequently adapts to these stages:

4.3.1 Launching Phase

During the first project stage, the prominent goal is to acquire a high number of first users that are interested in the Ridesharing Club. Here, the main activities concentrate on promotion activities that are continuously evaluated and monitored. In this context, it is important to analyze the main motivations of the lead users. During the launching phase, members can only access the internet-based ridematching service using the trip calendar. The main research question during the launching phase is how to efficiently acquire a critical mass of users under the condition that the main incentive of using a ridesharing service is that this critical mass and consequently a broad offer of lifts and potential passengers already exists. During this phase, it is also important to keep the lead users constantly informed on the service activities in order to keep their interest—even if chances for a successful ridematching are still small. During the launching phase, the evaluation focusses on expectancies and motivations of the first lead users who are generally interested in the service and willing to use it.

In order to analyze the promotion activities, the CARRIVA promotion staff documents all activities by weekly reports, short interview records with prospective customers and within group discussions. These documentations are the basis for monitoring and adapting the promotion strategy according to the target groups and locations for concerted activities. From the documentations, we can moreover conclude first results on user motivations and demands. The first "ex-ante" inquiry of lead users on motivations and opinions on the promotion activities is planned when 500 users have registered as members of the CARRIVA Ridesharing Club, while the second inquiry will be made when the club has reached a number of 1,000 members.

4.3.2 Test Phase

During the test phase, both the internet-based and the phone-based ridematching system should be tested by lead users. Here, first spontaneously calling "flexible users" will be matched with "permanent users".We evaluate the first experiences with CARRIVA in interviews and group discussions in order to determine expectancies and demands particularly with regards to the phone-based ridematching.

During the test phase, we aim to find out how members indeed use both central ridematching options during operation. The users will be intensively informed and accompanied during the introduction of the phone-based service, since this offer is much more unfamiliar than the internet-based ridematching service. For the evaluation, we analyze the communication between members as well as between members and the CARRIVA staff. Furthermore, matching activities and phone inquiries are documented by user protocols, which are also analyzed according to frequency, regional clusters and calling times.

The objective of the test phase is to optimize both ridematching systems according to their usability and technical operations. This also provides a basis for assessing actual potentials for the future normal operation. In this context, the main determinants are the main routes and travelling times in order to find convergences. For the evaluation, we use personal data from the registration for determining starting point and destination of the route, data from the trip calendar if available (passenger, driver or both, travel times of "permanent users") and data from an online-inquiry on actual shared trips and the according matching option (internet or phone). Finally, we interview certain lead users and organize a lead user discussion. This data provides a basis for evaluating the user acceptance of carpooling in detail, e.g. with regards to preconditions for ridesharing, external influences, spatial frameworks etc.

4.3.3 Expansion Phase

The results from the launching and the test phase will be the basis for evaluating the internet- and phone-based ridematching system and for determining the main supporting and restricting factors for carpooling services. For an outlook on the further expansion phase, we will analyze conditions that are particularly relevant for a further dissemination and a transfer on other ridematching services. Here, our interest focuses on the preconditions for matching spontaneous calls. Finally, we will assess sustainability effects of a ridematching service like CARRIVA.

During the expansion phase, we concentrate on frameworks and effects of dynamic ridesharing. On the example of CARRIVA, we develop recommendations on developing and organizing flexible carpooling services for commuters. Moreover, the overview on supporting and restricting frameworks, transfer options as well as on sustainability effects will help to support the idea of dynamic and flexible ridesharing in general

Chapter 5
Launching the Service: Reaching a Critical Mass by Establishing an Internet-Based Service

The objective of the first project phase was to reach a critical mass of users, since a carpooling service cannot continue without a minimum offer of trips and requests. As a first step, an internet-based service has been established to acquire a minimum number of users as the "backbone" of the service. Since an internet-database requires determining fix dates for travelling, only carpoolers with more or less fix travel times can use this service during the initial phase. Completing a personal profile with fixed travel times is simple and familiar for most user groups. Therefore, on this stage, there are barely general concerns on behalf of the users about the technological requirements of the service. Still, users have to show a general acceptance for the service concept and the technical components need to be user-friendly. There are, however, a high number of internet-based carpooling services, which do not meet this pre-requisition.

The main advantage of an internet-based matching system is that frustrating experiences are held to a minimum, since users only get positive feedbacks in terms of information on a successful match, while unsuccessful matches are not communicated. Therefore, an internet-based carpooling service is a useful first step for establishing a flexible carpooling service, since it does not provoke frustrating experiences due to unsuccessful matches. This is highly important for the launching phase of a new service, since frustrated users will not be willing to use this service again.

In summary, there are some basic conditions to consider in order to install a successful internet-based service and to reach a critical mass of users. In this context, we will firstly present relevant theories and evidences from different carpooling studies in Sect. 5.1. In Sect. 5.2, we will describe our activities during the launching process of the CARRIVA rider's club, before presenting the results of the accompanying empirical study in Sect. 5.3. Finally, we will evaluate the launching process as a basis for the following test and dissemination phase in Sect. 5.4.

V. Handke and H. Jonuschat, *Flexible Ridesharing*,
DOI: 10.1007/978-3-642-11345-1_5, © Springer-Verlag Berlin Heidelberg 2013

5.1 From Acceptance to the Actual Use of Internet-Based Carpooling Services

During the last decades, several scholars have developed psychological or economic models to explain the link between user acceptance of new business models and their success in the market. One central conclusion of nearly all models is that acceptance is a necessary, but not a sufficient condition for market success. If users appreciate a new product or service concept, they additionally have to have a strong motivation to use or buy it. Therefore, business innovations have to address both the acceptance and the motivation of the user. In the following section, we will therefore explain links and differences between both concepts as argued in scientific studies.

5.1.1 User Acceptance

User acceptance is mandatory for the success of new business models and has therefore become one of the central research questions in social and economic sciences (Kollmann 1998). Simon (2001) defines acceptance as a "positive agreement on behalf of the user. It can be defined as an antipode to the term of objection" (Simon 2001, p. 87, own translation). However, acceptance does not only consist of a unique decision for consumption, but also an emotional bonding to the product or service resulting in long-term consumption. There are a multitude of market focused studies on user acceptance that relate however to different research questions:

- *Market research studies*: Which are main consumers' requirements and needs with respect to the technical aspects of a product or service? Which are the main competitors and market niches? etc.
- *Studies on product and service characteristics*: Which product or service features enhance or impair user acceptance? Which characteristics are decisive for the user acceptance?
- *Marketing studies*: Which are the unique selling points of the product or service? Which are the best communication strategies for each target group?
- *Studies on customer relationships:* Which strategies result in a long-term customer binding? How can we strengthen the emotional binding to a product or service? etc.

All four aspects are decisive for a market success: While market research and studies on product or service characteristics are relevant for the design process, marketing strategies and customer relationships will be effective after launching the new product or service. During both phases, emotions for the product or service play an important role for user acceptance.

Academic research on acceptance often refers to introduction of social or business innovations. The first step to assess user acceptance is to analyze the emotional, cognitive and affective original conditions before introducing the innovation. Initial approvals and objections are important indicators for developing a promotion strategy for the launching phase. In this context, it is important to take concerns seriously, e.g. regarding the reliability of the service, technical processes or chances to get a matching partner, and to decisively address them, since emotions play an important role for the general acceptance of new and unfamiliar operations. On this basis, a SWOT analysis of strengths, weaknesses, opportunities, and threads can give a hint on first steps to optimize the service's performance and operations according to the emotional situation of prospective customers.

Müller-Böling and Müller (1986) distinguish two dimensions of emotional binding: Acceptance by attitude on the one hand and acceptance by behavior on the other hand (Müller-Böling and Müller 1986). Acceptance by attitude consists of an affective component that is based on emotional aspects, and of a cognitive component that derives from personal cost-benefits considerations. Acceptance by attitude is predominantly assessed by qualitative studies (Simon 2001; Kollmann 2000). In contrast, Müller-Böling and Müller (1986) state that acceptance by behavior are represented by the actual use of a product or service. Therefore, analyzing acceptance by behavior can refer to quantitative data on e.g. the number of actual users, frequency of use etc.

5.1.2 Motivation

Motivation is built on intrinsic processes preparing personal actions. According to the main German dictionary Duden, motivation is "the incitement determining human action" (Drosdowski et. al 1996, S. 505). The concept of motivation has been in the centre of psychological and sociological discussions, since philosophers of ancient Greece have debated on sources of motivation and action of human beings. During the second half of the 20th century, many scholars have added their perspectives from different perspectives to the discussion on motivation (e.g. Maslow 1954; Vroom 1964; Katz and Kahn 1978). In this context, Barbuto and Scholl (1998, 1999) have determined two intrinsic and three extrinsic sources of motivation (Barbuto 2002, p. 43):

- *Intrinsic process motivation* refers to personal pleasure the as driving motive for action.
- *Self-concept internal motivation* is based on personal values and guidelines representing with the ideal self, like e.g. efficiency or the will to change a situation,
- *Instrumental motivation* derives from monetary or other external incentives (e.g. power) that are associated with a certain action.

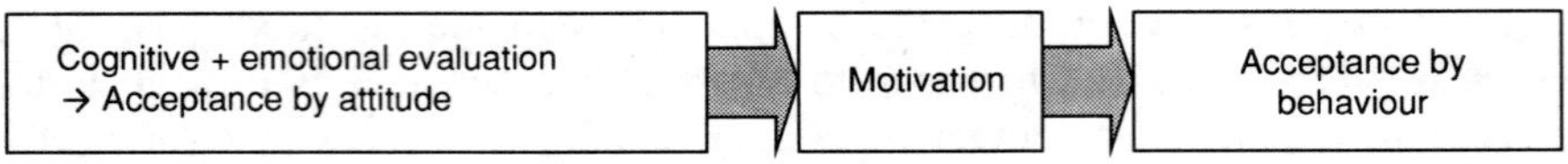

Fig. 5.1 Relation between motivation and acceptance

- *Self-concept external motivation* rewards followers with status, sense of belonging or a good reputation, when they meet social expectations within the personal environment.
- *Goal internalization motivation* means to adopt values and objectives of an external organization (e.g. employing company, associations etc.) and integrate them the personal value system.

Both intrinsic and extrinsic motives decide, if a general acceptance by attitude deriving from cognitive and emotional evaluation processes will result in action that manifests in an acceptance by behavior. The following figure illustrates this simplified acceptance model leading our evaluation (Fig. 5.1).

Applied to ridesharing services, this correlation implies that users firstly must have a positive attitude about the service resulting from a cognitive cost-benefit calculation and an emotional, mostly normative appreciation of the service concept. Beyond that, users must be motivated to use the service, which means to feel sufficient internal (e.g. an endeavor to be a "pioneer") and external motivations (e.g. reduced mobility costs) for actually sharing a ride. Evaluating the ridesharers' acceptance and motivation therefore needs to take these interrelations into account in order to assess the service's performance.

In this context, Rogers' (1954) innovation adopter categories link the psychological approach to user motivation with a macro-sociological perspective on the diffusion of innovations and their market effects (Rogers 1954, p. 22, see Fig. 5.2).

These categories illustrate that even in case of high general acceptance only about one out of six potential customers will actually use a service or a product during the launching phase. Innovators and Early Adopters usually show a high tolerance to a lack of service performance during the launching phase, but the majority of customers will not join in until the service is established and operations work error-free. As a consequence, it is important to optimize the service performance and operations as soon as possible in order to reach the Early Majority customer group.

Summarizing, motivation psychology theories as well as Rogers' theory on diffusion processes can help to understand the factors that lead from a general acceptance to the actual use of an innovative service. Yet, ridesharing is closely connected to daily routines and organizational frameworks that strongly influence and limit the range of personal decisions e.g. to work later or to vary the travel route. In this context, Ozanne and Mollenkopf (1999) have empirically developed a model that explains the personal decision process to carpool.

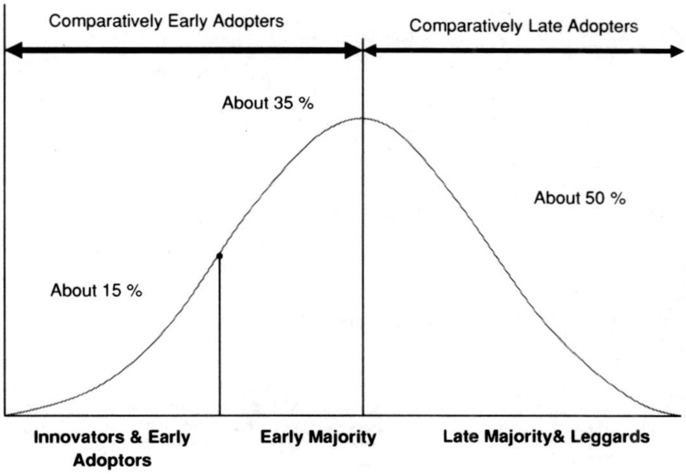

Fig. 5.2 Adopter categories for the diffusion of innovations

5.1.3 Ozanne's and Mollenkopf's Carpooling Behavior Model

During the last decade, several studies have analysed carpooling routines and behaviour in order to support and disseminate existing schemes. Ozanne and Mollenkopf (1999) distinguish four different categories of carpooling studies according to the main determinant considered:

1. Socio-economic factors like age, gender, income, employer or working times,
2. Route characteristics like distances, travel times and costs for commuting,
3. General customer demands like reliability or convenience, as well as
4. Personal motivations like the wish for new contacts

Some studies (e.g. Bonsall et al. 1984; Levin and Gray 1979) focus on the interrelation between personal, predominantly financial benefits, but also deal with the influence of social drivers like sustainable development goals or traffic jam prevention. Other studies analyze how to arrange carpooling and work conditions (e.g. Cervero und Griesenbeck 1988; Ferguson 1990). Declaring a lack of systematic approaches and theoretical concepts, Ozanne and Mollenkopf develop an own carpooling behavior model on basis of marketing theories, i.e. the TRA—Theory of Reasoned Action (Sparks and Shepperd 1992). and the TPB—Theory of Planned Behavior (Rogers 1954)

In this context, the TRA explains the relationship between attitude and behavior by a causal sequence from beliefs over subjective norms and attitudes to a behavioral intention and finally to action (Sparks and Shepperd 1992). The following figure illustrates this causal sequence (Ozanne and Mollenkopf 1999, p. 2, see Fig. 5.3).

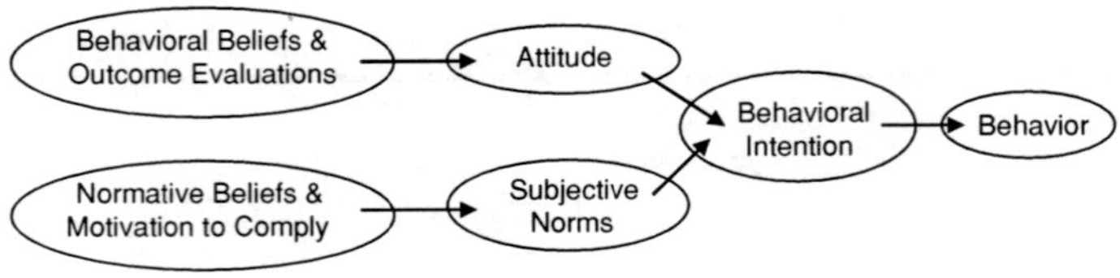

Fig. 5.3 Model of the theory of reasoned action (*TRA*)

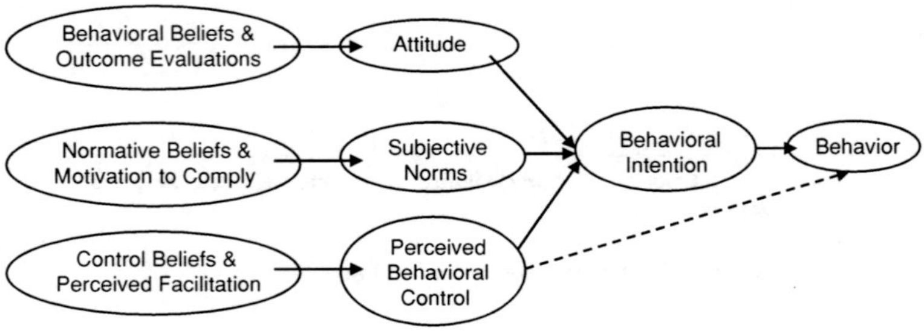

Fig. 5.4 Model of the theory of planned behavior (*TPB*)

Since the TRA only refers to situations when a person can completely control the individual's behavior, Ozanne and Mollenkopf integrate the Theory of Planned Behaviour (Rogers 1954; Taylor and Todd 1997) into their model (Fig. 5.4).

Here, a behavioral intention results not only from a personal attitude and subjective norms, but also from a perceived behavioral control that is based on control beliefs and a perceived facilitation. In addition, Ozanne and Mollenkopf (1999) have integrated past experiences as determinant for carpooling behavior in their survey on carpooling behavior. As a result, Ozanne and Mollenkopf (1999) found out that the intention to carpool is strongly related to the attitude of the respondents, which is determined both by personal relative advantages (e.g. to save money, to interact with others or to find a parking lot) and by compatibility to daily routines and commuting patterns (Ozanne and Mollenkopf 1999, p. 5). In addition, facilitating resources such as an existing ridesharing service has a direct impact on the attitude towards carpooling on the one hand and to the perceived behavioral control on the other hand. Finally, past experiences were found to have a positive effect on the willingness to carpool in future. In contrast to the TPB model, the empirical study could not find an evidence that subjective norms had an actual effect on carpooling behavior (Ozanne and Mollenkopf 1999, p. 7).

Summarizing, the most import influences on carpooling are personal advantages, compatibility, facilitating resources as well as past experiences (according to Ozanne and Mollenkopf 1999, p. 4, see Fig. 5.5).

Since ridesharing is mostly a new concept for employees at the Frankfurt airport, past experiences will assumingly not have a considerable effect on

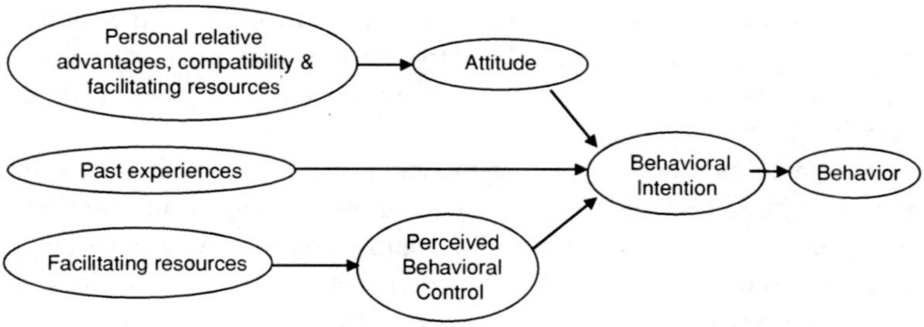

Fig. 5.5 Carpooling behavior as explained by the TPB

carpooling behavior. Consequently, personal attitude and perceived behavioral control will be the most important factors for assessing motivations to carpool in our research context. Apart from personal advantages and compatibility to daily routines, the ridesharing service as a facilitating resource has a twofold effect: On the one hand, it influences the personal attitude by representing a certain image (e.g. non-profit, environmentally friendly etc.), and on the other hand, the service directly determines the actual carpooling behavior by e.g. informing on ridesharing opportunities, organizing the matching or billing process.

5.1.4 Conclusions for the Launching Phase

All presented theories on acceptance, motivation and actual carpooling behavior document that normative beliefs and a general acceptance are a necessary condition, but do not directly result in using ridesharing services. Actually, carpooling behavior is rather dependent on the service's image and organization on the one hand and personal advantages as well as a compatibility to daily routines on the other hand. In accordance to Roger's (1954) theory on diffusion of innovations, most people are more open to carpooling, if they have experiences with ridesharing. Thus, during the launching phase, we can expect that only about 2.5 % of the potential customer group, the "innovators", will test the service. Related to a total number of 70,000 employees, the maximum theoretical number of lead users during the launching phase is only 1.750, and we can expect that the actual number of users will only be a fraction of this. It is therefore vital to optimize the service as soon as possible in order to teach the following consumer group of "early adopters".

For this reason, the service's image and organization has to strictly adapt to the customers' expectancies and to their daily routines. Firstly, all prospective users have to clearly understand all personal advantages of ridesharing, its compatibility to daily routines as well as the service's contribution to "greener mobility" in order to develop a positive attitude towards the service. Secondly, the service indeed has to respond to all customer reactions indicating a mismatch to daily routines or the perceived behavioral control in terms of e.g. restrictions to the personal commuting patterns, technical problems of the matching process, communication problems etc.

Marketing activities as prevalent canvassing strategy can positively influence the personal attitude towards a service. In contrast, technical and process-related adaption to user's routines requires developing completely new strategies e.g. on constantly monitoring customers' expectancies during the launching process, dynamically optimizing the service's features, integrating individual expectancies in the service concept etc. This is a challenging task, but an important success factor for dynamic ridesharing services that always have to adapt to regional frameworks and the target group's preferences. Thus, the main goal of the assessment activities is to detect central determinants not only for the user acceptance, but also for the motivation to actually use ridesharing services at the example of the CARRIVA Ridesharing Club.

5.2 Activities During the Launching Phase

The following chapter presents relevant results from assessing the CARRIVA Ridesharing Club launching phase in 2008. In January 2008, the launching phase has started with broad marketing activities at the airport in order to inform on the CARRIVA offer and to encourage memberships. Five months later, we have intensified the promotion activities by offering membership registration on the spot and by calling members in order to complete their profile and trip calendar.

5.2.1 Marketing Activities for Encouraging Memberships

In order to encourage employees at the Frankfurt airport to register at the CARRIVA Ridesharing Club, the promotion team has broadly distributed CARRIVA ball pens, CARRIVA leaflets and a hand-out explaining the registration and the ridematching process. Simultaneously, the promotion team has evaluated adequate locations and times for their activities. Furthermore, they have observed and documented their activities and personal contacts during the whole promotion phase. The following table summarizes the main information from these "observation journals" (Table 5.1):

The promotion team has left information leaflets to take away at some locations like gas stations, shops or pay machines on parking lots in order to reach all employees passing by. The following sections summarize the main outcomes of the promotion activities' evaluation.

Table 5.1 Summary of the observation journals

Location	Opening hours	Rush hours	Visitors	Notices
Terminal shops	n.a.	Afternoon, evening	Shop assistants	Positive reactions, but first information has been sufficient.
Canteens				
1	n.a.	10:45 a.m.–2:30 p.m. und 5:00–7:00 p.m.	Operative staff, few "white collars"	Interested, but most people are busy
2	Mo–Fr	Lunchtime and after 5 p.m.	Mixed staff	Lift has been a good location
3	n.a.	7–9 a.m., 11:30 a.m.–2:00 p.m. 5:00–7:00 p.m.	Mostly "white collar staff"	Very good location
4	n.a.	12–14 p.m.	Mostly operative staff and truck drivers	Good location, a lot interest
5	n.a.	7–9:30 Uhr, 10–1:30 p.m., 5:15–7:30 p.m. (am WE eher weniger)		Good location, a lot interest
6	Mo-Fr		"White collar staff", higher management	Less interested
7	Mo–Fr until 1.30 p.m.			Good location, a lot interest
8	Mo–Fr bis 3 p.m.		Mostly "white collar staff"	Good location, a lot interest, interviews with smokers
9	Mo–Fr. until 2:45 p.m.	11 a.m.–2 p.m.	Mostly "white collar staff"	Good location, a lot interest, nice place
10	6.30 a.m.–4 p.m.	9 a.m.–2:30 p.m.; 4–7:30 p.m.	Operative staff	Unconfortable location
Parking lots				
P 1	n.a.	2:30–3 p.m., also weekend	Prefield workers and office staff	Good location at the entrance and the bus station, but busy employees (handouts!)
P 2	n.a.	no	Mixed	Interested employees
P 3	n.a.	8–10 a.m.	Mostly "white collar staff"	Interested, but busy
P 4	n.a.	1:30–3 p.m.	Mostly operative, but also "white collar staff"	Interested, but busy
P 5	n.a.	1–3:30 p.m.	Mixed	Good location, a lot interest

5.2.1.1 Effectiveness: Awareness of the CARRIVA Ridesharing Club

During the five months of promoting activities, CARRIVA has averagely gained 50–60 new members per week (cp. Fig. 6). The observation journals revealed that respondents have not adequately noticed the promotion activities, and particularly posters have not been remarked by people passing by.

The observation journals have revealed that some respondents have not noticed the promotion team and suggested to support the campaign by further posters. However, it turned out to be difficult to find suitable places for the posters. The journals documented that after three months the name "CARRIVA" was widely known, because more than half of the addressed persons had already seen a poster or a flyer.

5.2.1.2 Good Locations for Promotion Activities

Although the plan was to promote CARRIVA primarily on car parks in order to directly address drivers, it turned out that on the car parks, the promotion team could only give brief information or hand over of a flyer. The reason for this is that interviewees are usually in a hurry on car park areas and additionally parking lots seemed to be inconvenient locations for a chat. Within the parking areas, particularly lifts have been good locations to address passers-by, since they represent a bottleneck that the drivers have to walk by.

Finally, canteens and cafeterias turned out to be the best locations for promotion activities, since prospective members can get first information before lunch and ask questions after the meal. Besides, this can initiate conversations on the CARRIVA concept with other employees at the table. Moreover, canteens are regularly used and potentially interested people can get further information or information material during the following days. It has been particularly effective to distribute handouts on the tables in the canteens. Nevertheless, even in busy canteens, only few people took their time to ask for information on CARRIVA at peak hours.

After the first weeks of the promotion campaign, various giveaways have been distributed at the CARRIVA information desks, like e.g. candies or ballpens. This had the effect that employees have actively approached the promotion staff and asked for further information. Furthermore, owners and staff of the terminal shops have been informed on the Ridesharing Club.

The promotion staff has remarked that particularly visitors of canteens for technical and operative staff have been interested in the CARRIVA concept. In contrast, promotion activities in canteens for managers and office workers have not been very responsive.

5.2.1.3 Special Promotion Activities: Using Corporate Information Channels and Corporate Events

In addition to direct activities on site, corporate information channels have been used to disseminate information on CARRIVA:

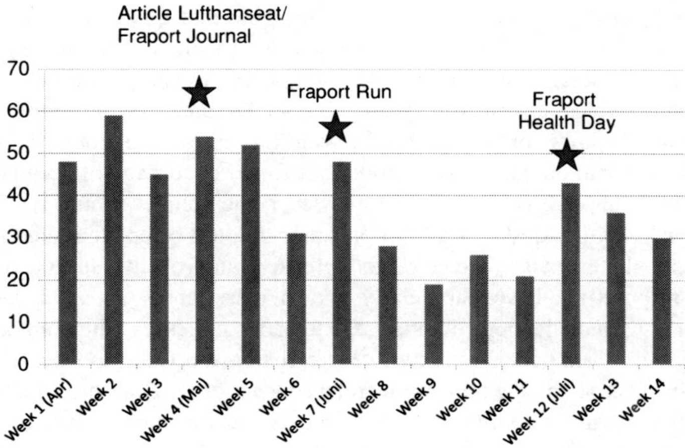

Fig. 5.6 Numbers of new registrations per week

- A posting in Fraport in-house journal,
- Articles and postings about CARRIVA in Lufthansa's corporate in-house journal and in Lufthansa's internal online portal,
- Direct link to CARRIVA web-site from Lufthansa's internal mobility web-site.
- A CARRIVA supplement in Lufthansa's corporate journal "Lufthanseat".
- Information desk at the "Fraport Run" and on Lufthansa Cargo´s Climate Care event.
- Distribution of handouts on the "Fraport Health Day".

All of these activities have resulted in a subsequent increase of daily registrations. In total, over 800 persons have registered as users within 5 months. Compared to other ridesharing services this is a high number of registered persons, especially when considering that at the start of the promotion campaign, the phone-based ridematching option as a unique selling point of the Ridesharing Club has not been available yet. Figure 5.6 illustrate the correlation between new registrations and special promotion activities.

After the first two months, new registrations have steadily decreased in spite of intensified promotion campaigns, which can be explained by a general saturation effect.

5.2.1.4 Reservations About CARRIVA

The observation journals have revealed that some people have also expressed reservations about CARRIVA, since they could not completely understand the concept of internet- or phone-based ridematching. These respondents have not experienced ridesharing yet and could not imagine that this could actually work. It is important to take these reservations serious in order to approach also customer groups that are unfamiliar with alternative transport options. Therefore, ridesharing services also need "low-threshold" strategies for people who are either not media-savvy or just not used to ridesharing. For this rather sceptical target group, personal contact is very important to get encouraged to register for the Ridesharing Club and use the service. However, the contact should rather be passive, e.g. information desks or a high presence due to posters and handouts, since sceptical people have also reacted rather defensive on proactive approaches.

Some respondents have particularly shown reservations about declaring their email addresses and telephone numbers due to bad experiences with commercial promotion activities, and poor data security. Here, firstly reserved respondents have been particularly appeased by the non-commercial concept of the Ridesharing Club as well as by the trustworthy supporters of the project. The legitimation on behalf of Fraport, Lufthansa and the Federal Ministry of Transport, Building and Urban Development has been in general a strong "door-opener" not limited to sceptical groups.

In addition to these general concerns, some of respondents have expressed reservations concerning the CARRIVA concept, since they doubted the success of the service. These respondents have, however, been open to the argumentation that the membership in the Ridesharing Club is for free and that the project is still in the development phase. Here, the promotion staff has remarked that a transparent communication also on problems during the starting phase of the non-profit ridesharing service has particularly been convincing sceptical users, whereas strongly advertising the advantages of ridesharing could result in even more reservations within this user group.

5.2.1.5 Ridematching Demonstration on Mobile Stations

After the first weeks of the promotion campaign, the number of daily registrations has been considerably decreased. Thus, a main aim of further promotion activities has been to signalize that CARRIVA is still active and constantly optimizing its offer. From the provider's perspective, members should be moreover encouraged to also use the trip calendar. For this reason, mobile laptops were used on site on the one hand, and certain members have been directly called on the other hand. Finally members who had entered their travelling times in their trip calendar have got an e-mail with their matching results.

Laptops with a mobile internet connection have been used as mobile stations to encourage interested persons to register for the Ridesharing Club as well as to use the currently available service, i.e. the matching over the trip calendar. With the help of the mobile stations, the promotion staff could demonstrate the registration and matching process on site. The mobile stations also aimed to demonstrate the new geographic visualization of locations and driving routes.

As a low-threshold access, the personal assistance for registration and the use of the web-based ridematching process has particularly addressed users with

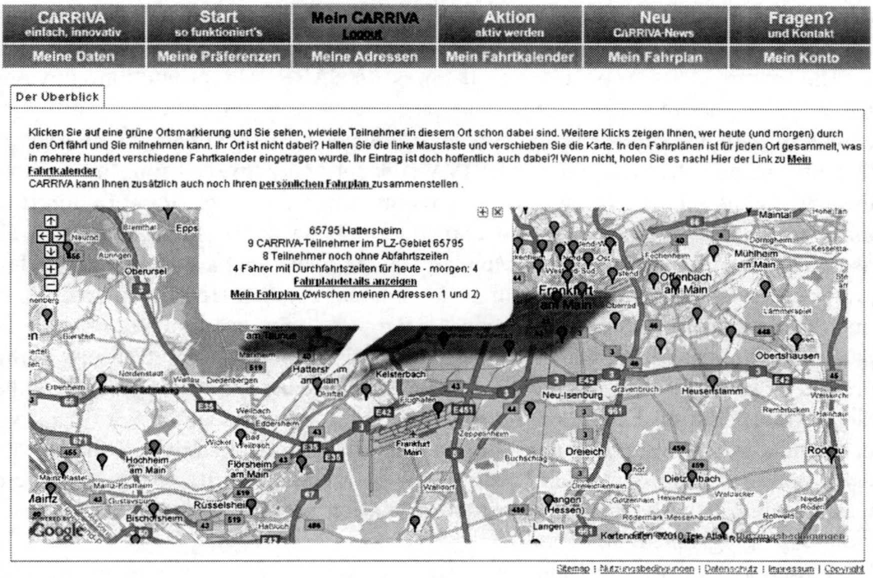

Fig. 5.7 Query for matching partners

reservations about new technologies. While the registration process did not pose any problems, several persons have expressed a need for support with regards to the use of the trip calendar.

For the demonstration of the matching process, the CARRIVA homepage has been expanded by a query tool that shows all registered members on a map per postal code area. Here, users could enter a certain postal code and get results on the total number of registered members within this area and their driving times. A link leads each user to the timetable for this route (screenshot of the CARRIVA homepage 2008, see Fig. 5.7).

Whereas the respondents particularly appreciated the query tool, they have only shown interest in mobile station demonstrations at certain events like the Fraport Run, but not on general workdays.

5.2.2 *Results on Promotion Activities During the Development Phase*

A high number of web-based ridesharing services have been founded by non-profit initiatives. Therefore, a lot of initiatives like the CARRIVA Ridesharing Club do not have the financial resources to plan a professional roll-out with intensive marketing activities. With regards to the high importance of promotion activities for the success of services, many non-profit initiatives need to find an adequate balance between costs and benefits in order to gain as much attention as possible.

The empirical study on the effectiveness of CARRIVAs promotion activities has revealed that the cost-benefit ratio can considerably differ between different activities. Although, for instance, personal contacts are very effective with regards to both raising interest in the CARRIVA concept and actively using the service, they also bind a lot of resources for planning and on-site activities. In contrast, mass mailings are impersonal and anonymous, but they could indeed reach all users with a minimum of efforts. Moreover, some members have also expressed to prefer an anonymous information mail to personal contact. Here, corporate information channels like corporate journals or intranet communication seems to be of particular interest for an effective dissemination among the target group. Being present on certain events has been a very effective means to personally address the target group and raise attention. At the same time, expenses in terms of personal or financial resources are manageable. In fact, the empirical study has shown that using corporate information and events have been the most effective and efficient way to interest the target group in the ridesharing service, documented in direct increases of registered members.

Non-profit services should moreover actively use their non-commercial approach for promotion activities. In contrast to commercial mobility services, the respondents have been expressed a particularly trust in both non-profit services and research activities. Here, they are willing to accept definitely more deficiencies during the launching phase than with regards to commercial service offers. For enhancing the trust in the service, it is therefore important to give transparent information on the development process, and to constantly integrate the users into this process.

5.3 Survey Results on User Acceptance of Ridesharing

In order to evaluate the user acceptance of ridesharing and the motivation to use ridesharing services, we have analyzed a broad data set, from registration data and tip calendar entries to matching protocols and observation journals. Furthermore, we have used data from two surveys.

In May 2008, we have personally interviewed 287 airport employees on their acceptance by attitude (cp. Sect. 5.1) of ridesharing in general. At that time, about 360 people have already registered for CARRIVA, but also employees who have not yet registered have participated in this survey. The results from this survey are summarized in Sect. 5.3.1.

The second survey has focused on the acceptance by behavior (cp. Sect. 5.1), including only registered members. This survey has been conducted in July 2008, and 95 people, i.e. about 10 % of all Ridesharing Club members at that time, have answered the questionnaire online. Within the online survey, respondents could rate the usability of the homepage and bring in their experiences with the web-based ridematching system. Main results from the second survey are summarized in Sect. 5.3.2.

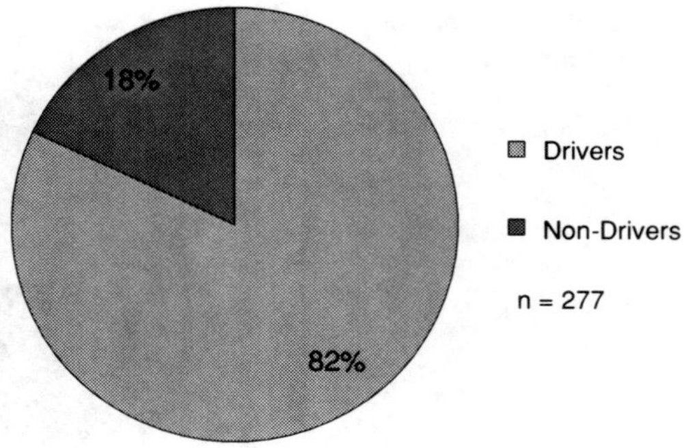

Fig. 5.8 Relation drivers-passengers

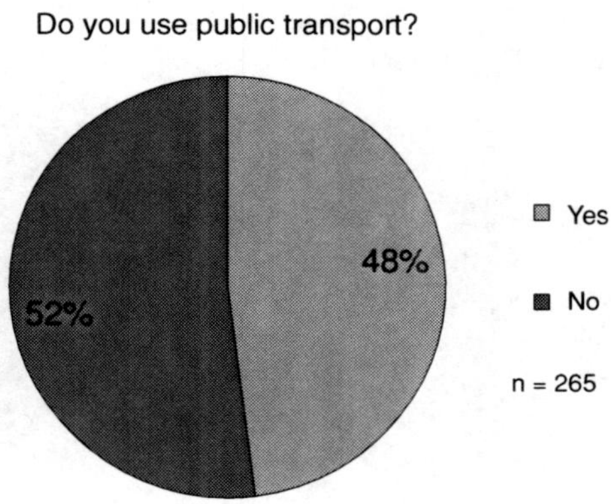

Fig. 5.9 Public transport user

5.3.1 Results on the Acceptance by Attitude of CARRIVA

The first survey could give insight into the users attitudes towards ridesharing, since they have not experienced ridematching processes or the CARRIVA offer yet and could therefore only give statements on their general opinion. The interviewees have shown a general interest in the concept and consequently represent lead users who are not familiar with flexible ridesharing services yet.

The first is a large share of drivers, since about four out of five respondents were car drivers (Fig. 5.8).

Unless at least half of the drivers are not willing to take a lift, the sample does not comply with the optimal driver-passenger ratio of 1:2 (cp. Sect. 3.3.20). However, some of the drivers obviously also use public transport, since half of the interviewees use public transport as well (Fig. 5.9).

In total, about a third of the respondents are both public transport users and car drivers. This group is particularly important for ridesharing services, since they apparently have different transport alternatives and could alternately be a driver or a passenger. However, half of the respondents never use public transport at all, and

Fig. 5.10 Frequency of public transport use

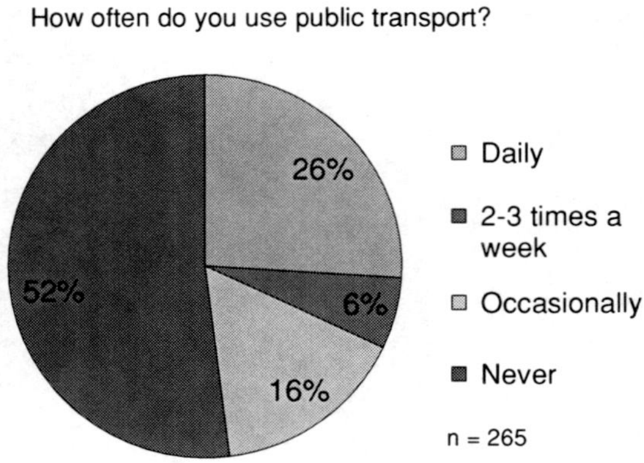

Fig. 5.11 Interest in using CARRIVA

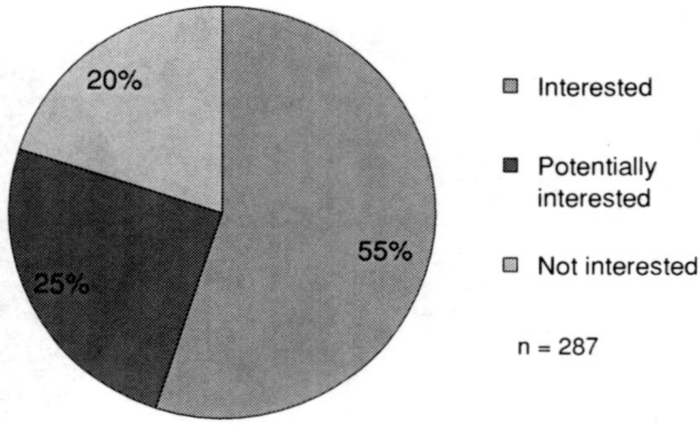

they could only give a lift to other Ridesharing Club members. For estimations on environmental effects of ridesharing, it is particularly important to assess motivations and acceptance structures of this group of "convinced drivers", since changing their opinion will have a considerable effect on the total number of car rides (Fig. 5.10).

More than three quarters of all 265 interviewees have shown a general interest in using CARRIVA in future, whereas one out of five did not show any further interest in the Ridesharing Club (Fig. 5.11).

Still, the interest is closely connected to gender, work, age and working scheme. The following data only includes respondents who are generally interested in the CARRIVA service. Particularly the gender seems to play an important role for the acceptance of ridesharing (Fig. 5.12).

Nearly 90 % of all interviewees who have shown interest in the CARRIVA concept are men. This is enforced by the fact that usually women are overrepresented in general surveys. With regards to the age of interested respondents, two thirds are under 40 years old (Fig. 5.13).

More than a half of all interested respondents are between 30 and 40 years old, whereas only a quarter is 40–60 years old. With respect to the working area, operative staff amounts to more than 40 % while about a third works in administration (Fig. 5.14).

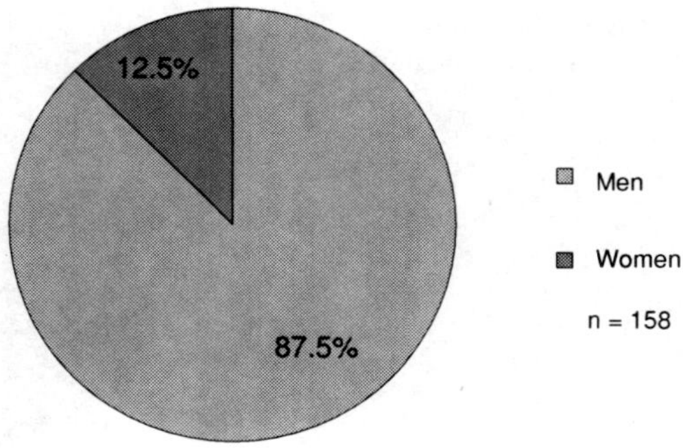

Fig. 5.12 Interest in CARRIVA according to gender

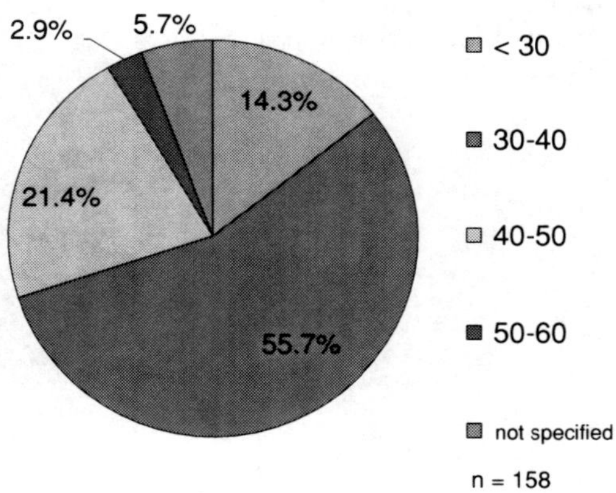

Fig. 5.13 Interest in CARRIVA according to age

Operative staff works predominantly, but not exclusively in shifts. This is also reflected by the fact that more than half of the interested respondents work in shifts (Fig. 5.15).

Only a quarter of interested interviewees work in flexible work schemes, although they are the main target group for flexible ridesharing services like CARRIVA. Still, nearly 18 % of the respondents neither worked in shift nor in flex-time models.

Regarding the interest in CARRIVA according to the mobility type, car drivers are very dominant (Fig. 5.16).

From the fact that 82 % of all interviewees, but 90 % of all interested interviewees are car drivers, and 42 % of public transport users are interested, while their total share in the survey is 48 %, we can conclude that CARRIVA predominantly addresses car drivers, while public transport users are less interested. Therefore, it is particularly important to intensify activities to promote the concept of ridesharing among potential passengers. However, more than 40 % is still a good share of interested users who obviously could also use public transport as alternative, e.g.

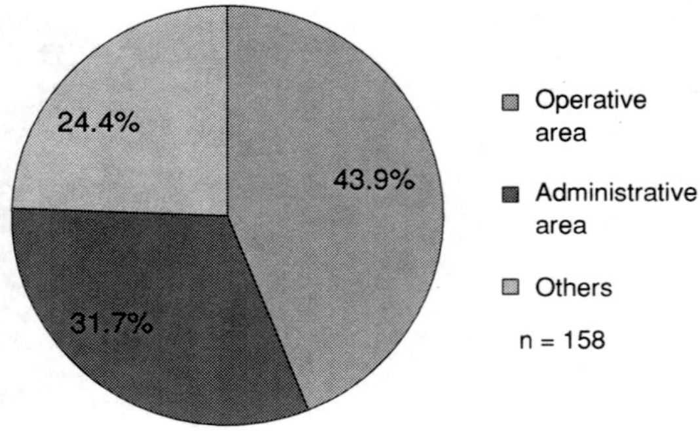

Fig. 5.14 Interest in CARRIVA according to working area

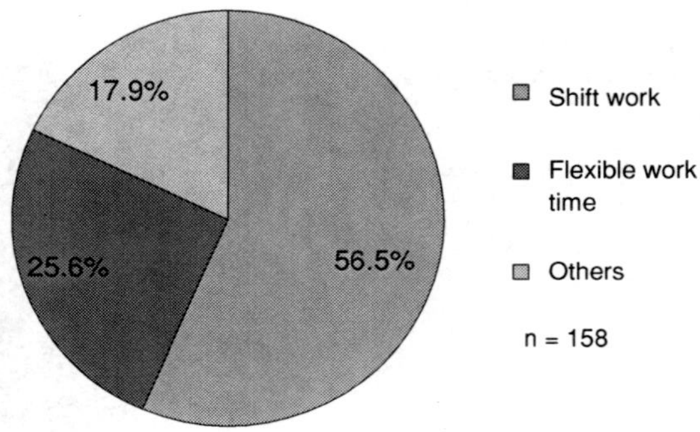

Fig. 5.15 Interest in CARRIVA according to work scheme

for the ride home. With regards to environmental effects, it is though even more important to convince car drivers to abstain from using their car for commuting.

With respect to arguments for an interest in CARRIVA, nearly nine out of ten respondents intend to save transport costs, while two thirds hope to contribute to environmental protection, and one third aims to save time (Fig. 5.17).

Here, car drivers particularly expect to save costs and contribute to environmental protection, whereas public transport users expect to save travelling time while commuting.

Some of the respondents who are not interested in ridesharing services at all have expressed general reservations that can be summarized in three clusters:

- *Criteria for exclusion*: "I don't have internet", "I'm a trucker and so I'm not allowed to give a lift", "I live right next to the airport".
- *Perceived as wrong target group*: "I work in shifts, this does not work", "I've got children that still go to the kindergarden", "We already have a private ride-sharing team", "I'm a pensioner and only like to have lunch here", "I only need to go to the airport two or three times a month".
- *Criticism of the concept*: "I don't like to make detours and that's a precondition for this service", "It's a good idea, but I don't believe that it could work".

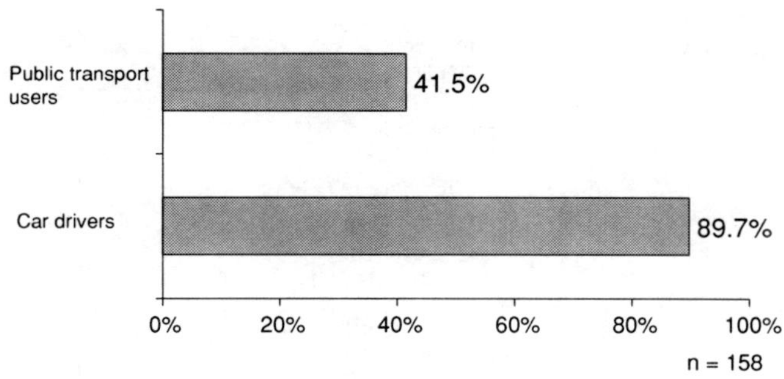

Fig. 5.16 Interest according to mobility type

Fig. 5.17 Reasons for interest in CARRIVA

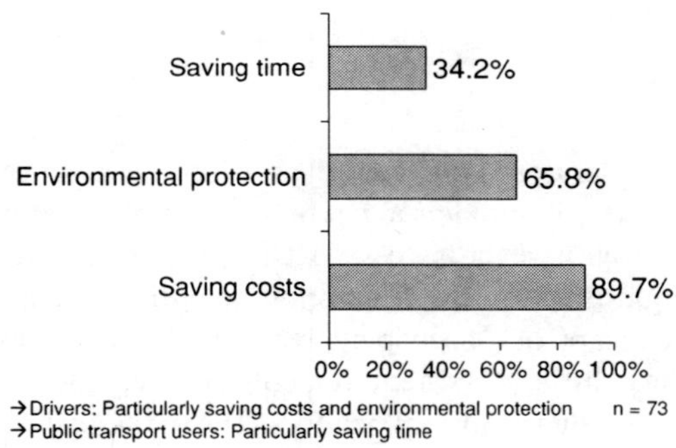

People who have no internet connection or who are not allowed to give lifts are indeed not the right target group for CARRIVA. In contrast, airport employees or visitors who do not perceive themselves as target group or do not trust in the success of ridesharing services can perhaps be convinced by further argumentation.

Summarizing, typical airport employees that are interested in ridesharing concepts like CARRIVA are car drivers between 30 and 50 years and particularly work as operative staff. Supplementary to the results on the attitude towards CARRIVA, the second survey has rather focused on the motivation to share a ride as well as on actual experiences.

5.3.2 *Results on Acceptance by Behavior*

The second online survey has addressed all 900 members at that time. About 10 % (95) members have completed the questionnaire. The survey's goal was to ask the first Ridesharing Club members for their user preferences and the performance of

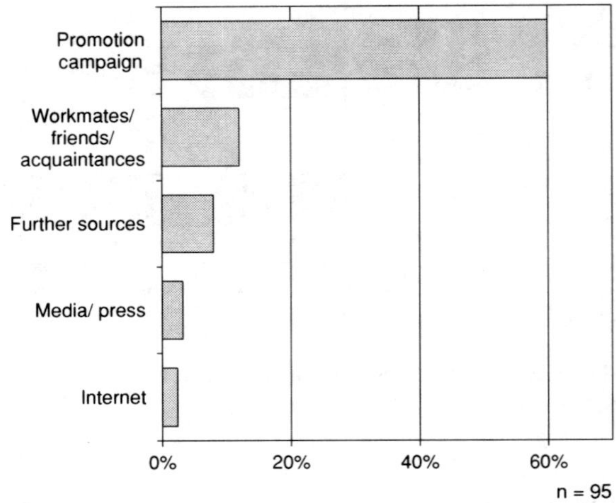

Fig. 5.18 Information sources for CARRIVA

the service. This data gives a hint on a general acceptance by behavior, although there still remains a gap between user acceptance and actual ridesharing patterns.

With regards to CARRIVA's performance, the online survey has contained questions on the promotion activities. We also tried to find out how the service concept has been perceived by the users. Furthermore, the survey should give insight in users' requirements and demands for ridesharing services as well as on individual mobility patterns. The main goal of the second survey has been to adapt the service's performance to customer demands.

5.3.2.1 Results on Promotion Activities

In order to detect main information channels for CARRIVA, we have asked for the first contact medium (Fig. 5.18).

Two thirds of all registered respondents have registered as members due to the promotion activities at the airport. Yet, one out of ten has heard of CARRIVA from acquaintances, friends and workmates, although the service has still been in the launching phase. This emphasizes the influence of social networks on strategies to gain new customers.

In addition, the respondents should assess the quality of different information media (Fig. 5.19).

The respondents have particularly emphasized the CARRIVA leaflet and the homepage as information medium as well as the personal encounters with the promotion staff. However, only 70 % of them have rated the leaflet and a little bit more than the half the homepage and the promotion activities as "very good". Information e-mails and articles have been rated as "very good" by about 40 %, and only one out of five have found the handout and the posters helpful. However,

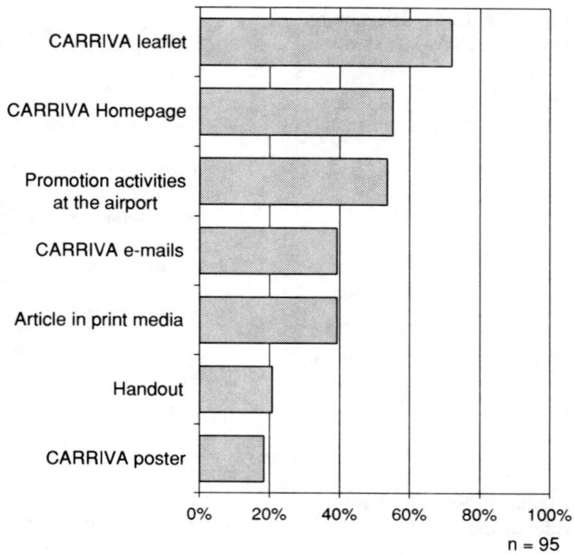

Fig. 5.19 Rating information material and media as "very good"

only the e-mails and the homepage have been definitely reached all members. Thus, a lower rate of the other media does not automatically correspond to a lower rating, but could also imply that the respondents have not got or noticed these sources.

Within the survey, the members have also been asked for their initial motivation to register for CARRIVA (Fig. 5.20).

Here, more than four out of five respondents hope to save fuel costs by ridesharing, and more than a half look for an alternative to public transport. Normative reasons like a contribution to environmental protection and to avoid traffic jams are also a quite popular motivation for half of the respondents. About a quarter referred to the trip itself by intending to be more independent from the car in general or by hoping for having a more interesting trip or gaining new social contacts.

In contrast, the members have also several reservations with regards to ridesharing (Fig. 5.21).

Two thirds fear not to find a matching partner and more of the half worry about finding an adequate agreement for sharing the ride. These reservations are particularly serious, since in fact, chances for a successful matching are still low during the launching phase. Therefore, positive experiences are an important incentive for initial members to gain trust in the service. In this context a high transparency on matching chances and information on "second best" matching results could diminish the risk of frustration due to a lack of matching success. This could also enhance the tolerance for initial problems in finding suitable partners.

About a third of the respondents relate their reservations to the ridesharing partner who could be unfriendly, or a risky driver. Finally, one out of four members

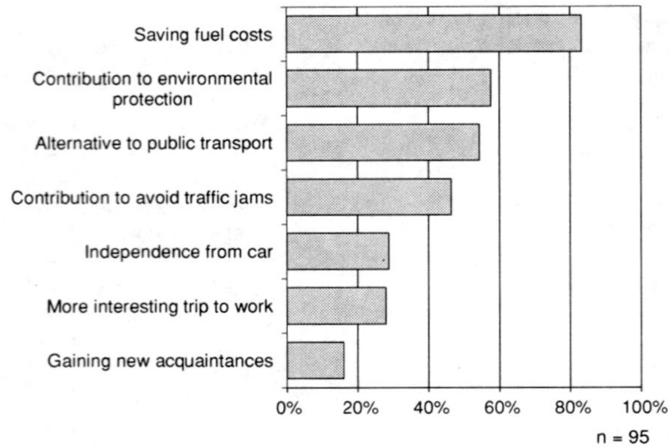

Fig. 5.20 Initial motivation to register for CARRIVA

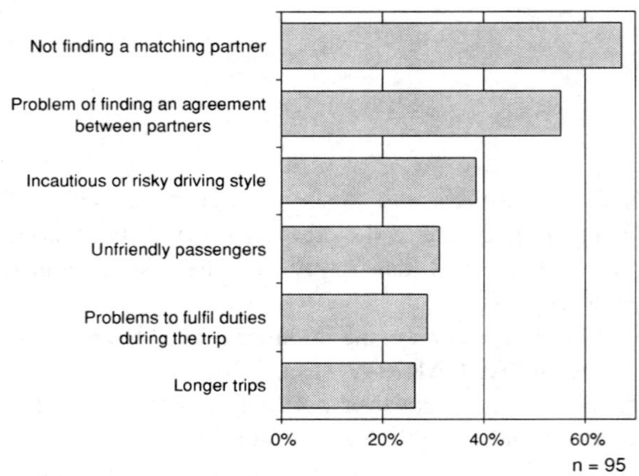

Fig. 5.21 Reservations on ridesharing

express concerns about their mobility routines in terms of longer or less flexible commuting trips.

We can assume that most of the reservations will diminish with the number of positive experiences both with the service or ridesharing in general. "Success stories" of other members could also be a means to illustrate that the service is already working for at least a part of the members.

5.3.2.2 Usability of the Matching Process

Phone based ridematching is a new and unique offer, and none of the users have similar experiences with that kind of ridematching. In order to enhance trust in the service, users consequently have to completely understand the communication

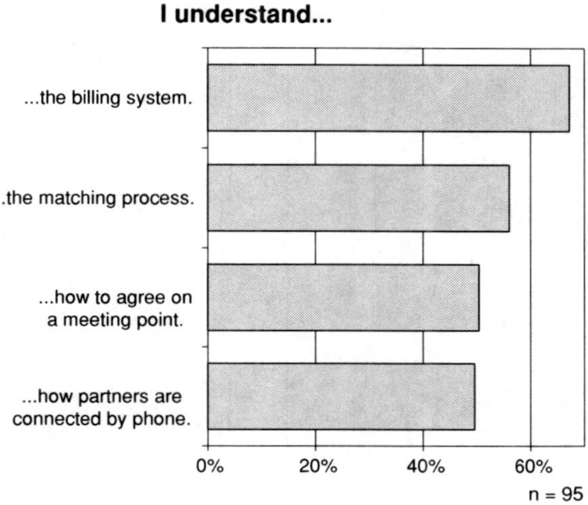

Fig. 5.22 Understanding the phone based matching system

processes during the matching process. However, the survey has revealed that there are still a number of information deficits (Fig. 5.22).

Two thirds of the respondents have not understood the billing system. Although this feature is not available yet, it will be a very important incentive to join the Ridesharing Club in future. Therefore, users should be informed on the billing systems already at an early stage of the service. Half of the respondents have also expressed that they neither understand the matching and the communication process nor the way to get an agreement about the meeting point. Here, it is also essential that potential users clearly understand how to actually use the system, since this is a necessary requirement for an active involvement in the ridesharing service. All information sources should therefore include user-friendly explanations of the actual matching process.

At the time of the survey, the internet-based matching system has already been active for several weeks. The survey has therefore also referred to reasons why members have not used the trip calendar yet (Fig. 5.23).

About one out of six of all respondents have stated that they are only interested in phone based and not in web based ridematching. More than 10 % have indicated that they cannot plan their trips in advance and that the trip calendar is therefore not interesting for them. Both reasons are exclusive criteria for web based ridematching on the one hand, but also show the need for flexible, phone based ridematching services on the other hand. In addition to this, 8 % of the respondents need further information on how to use the internet-based ridematching system.

In summary, we can conclude from the survey results that an efficient, stable and secure technical ridematching platform is a necessary condition for user acceptance, but appropriate and comprehensible information as well as simple user routines are the main success factors for ridesharing services. At the time of the survey, users still have however missed clear information on the system and

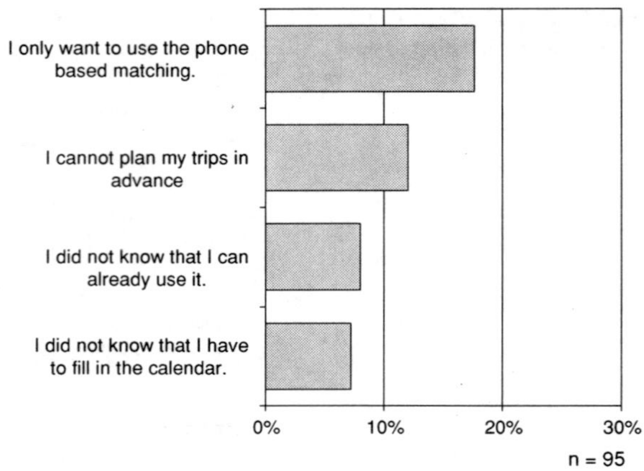

Fig. 5.23 Reasons for not using the trip calendar

did not really understand how to use it. This means for the dissemination phase that users demand up-to-date and comprehensible information as well as an intuitive navigation through the matching system.

5.3.2.3 User Preferences for Detours

The willingness to make a detour is an important success factor for finding a matching partner, since it directly enhances the route convergence: The longer the detours, the more ridematching partners will be found. Here, the general willingness to make detours is already considerably high (Fig. 5.24).

Only 3 % of all respondents are not willed to make a detour at all, whereas 20 % do not mind to make a detour on every journey. A further quarter would accept a detour, if they do not have further duties. One out of six would either exceptionally make a detour or make a detour for an extra fee. In total, three quarter of all respondents would explicitly make a detour in order to enhance the chance to find a matching partner. With regard to the distance of the detour, we also get a clear picture (Fig. 5.25).

According to this, two thirds of the respondents accept a detour of more than 5 min, and one out of five even accepts a detour of more than 10 min. In total, more than 90 % of all respondents are willing to make a detour of 5 min in order to pick-up a passenger. This actually relates usually to a detour of 4–5 km, which therefore is recommended as standard setting for the matching data base.

5.3.2.4 Information on Matching Results Before Departure

For the phone based matching system, it is important to determine the duration between the phone call for announcing a ride and the actual departure time. The CARRIVA system has calculated with a standard time of calling 20 min before

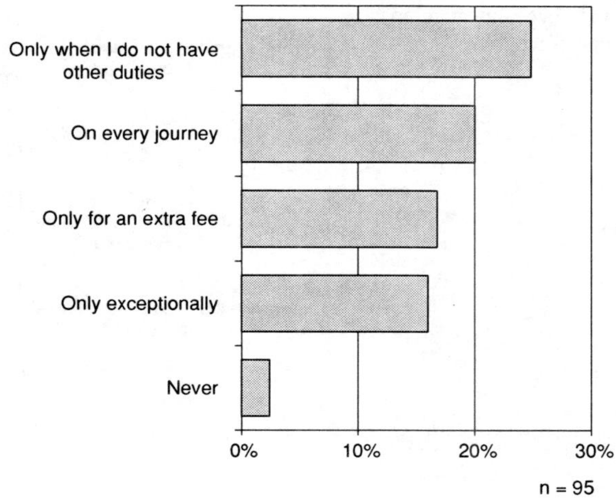

Fig. 5.24 General willingness to make a detour

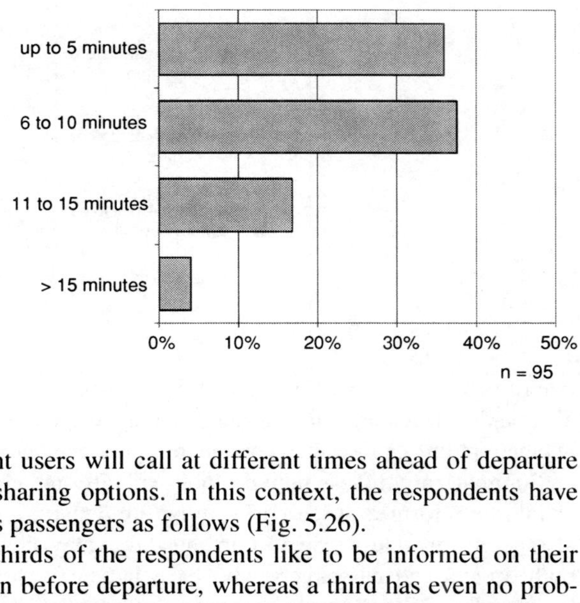

Fig. 5.25 Accepting a detour

departure. However, different users will call at different times ahead of departure in order to check their ridesharing options. In this context, the respondents have specified their preferences as passengers as follows (Fig. 5.26).

When taking a lift, two thirds of the respondents like to be informed on their matching results until 16 min before departure, whereas a third has even no problem to get a contact to the driver in less than 15 min before departure. Still, one out of six passengers need to get information on potential lifts at least 45 min before departure. Thus, one third plans their trip long-term, one third medium-term and one third short-term.

In contrast, drivers are much more spontaneous than passengers (Fig. 5.27).

More than three quarter of the respondents state that they need information on potential passengers half an hour before departing, and half of the respondents even like to be informed less than 15 min ahead. Still, about 12 % need to know 45 min

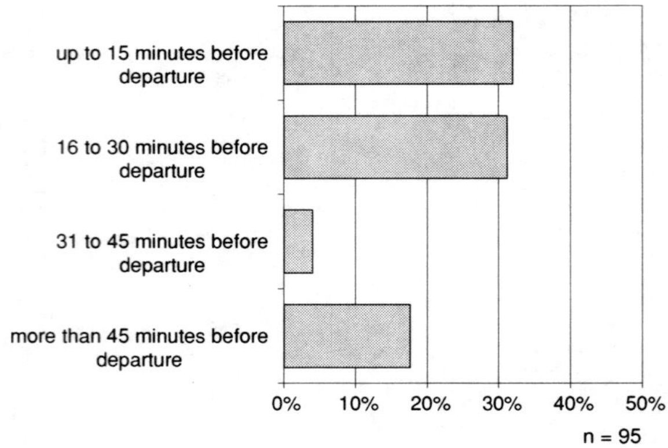

Fig. 5.26 Information on a matching driver

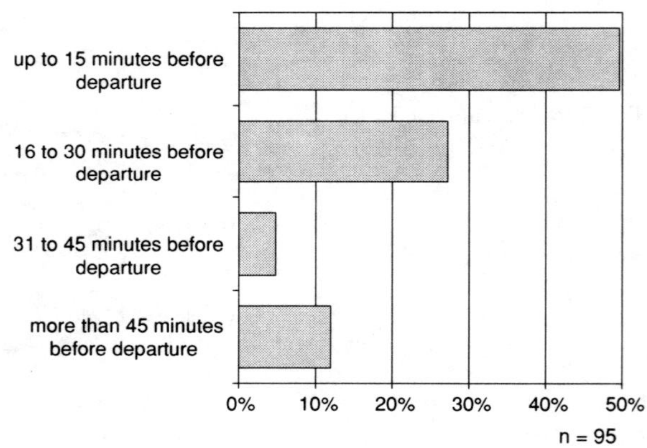

Fig. 5.27 Information on a matching passenger

before departure, if they will give a lift to one or more passengers or not. We can assume that depending on their ridematching result, these drivers would like to choose another option, e.g. taking the public transport, another route or departure time.

Summarizing, unless they did not explicitly indicated another time, passengers should be informed at least 15 min before departure on a lift, while drivers could also be informed less than 15 min ahead that they should pick-up a passenger. Still, both drivers and passengers who have indicated a time span for departure (e.g. between 8 and 9 a.m.) need to inform the system at least 15 min ahead on the exact departure time, if they leave earlier than the latest travelling option (e.g. 9 a.m.).

5.3.2.5 Mobility Patterns

For planning a ridematching service, it is helpful to analyze mobility patterns of the relevant target group. Here, average travelling times give a hint on general

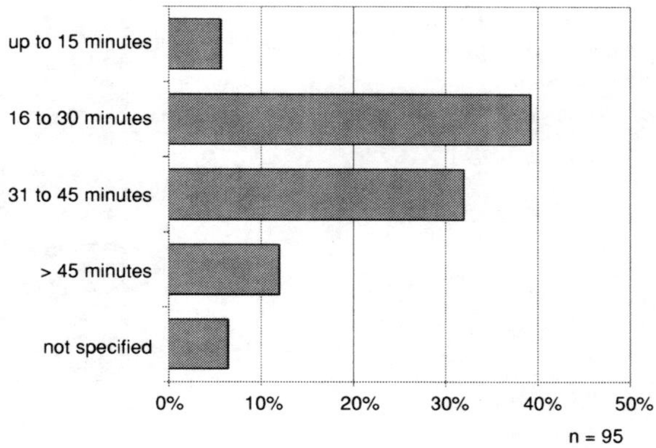

Fig. 5.28 Average travelling time to work as car driver

potentials for ridematching service within a certain area. In the area around the Frankfurt airport, commuters on average need less than 45 to get to work (Fig. 5.28).

More than three quarters of the respondents drive less than 45 min one way, and half of the respondents even need less than half an hour. Still, 12 % also commute more than 45 min to work. Public transport users averagely need much more time to get to work (Fig. 5.29).

Only 9 % of the respondents need less than 45 min to get to work by trains and busses. Half of them need more than one hour by public transport, one out of five even more than 1, 5 h. This is closely related to the high share of drivers among the survey sample (see Fig. 5.8), although we have to consider that still a half of all respondents also use public transport for commuting (see Fig. 5.9). Consequently, these figures reveal that ridesharing could be a good commuting alternative within the Frankfurt airport area, since in spite of a good public transport network and job tickets, travelling by train and busses seems to take much more time for commuters than driving by car. Given the fact that within the survey sample, more than 80 % are drivers, and about half of them also occasionally use public transport, the potential for taking a lift instead of taking their car is interesting for at least 40 %. This group has moreover also the option to take public transport e.g. on their ride back home, even if they potentially need much more time.

Giving a lift or being a passenger restricts the drivers flexibility, particularly when he or she uses to make stopovers on their way to or from work. However, only few drivers frequently make a stopover on their commuting trip (Fig. 5.30).

Yet, four out of five respondents never or seldom make a stopover, and while 12 % frequently stop for e.g. shopping or further duties. Therefore, only few members will experience less flexibility due to ridesharing. In any way, this group of "stopover-commuters" can also integrate their stopover addresses in their CARRIVA trip calendar.

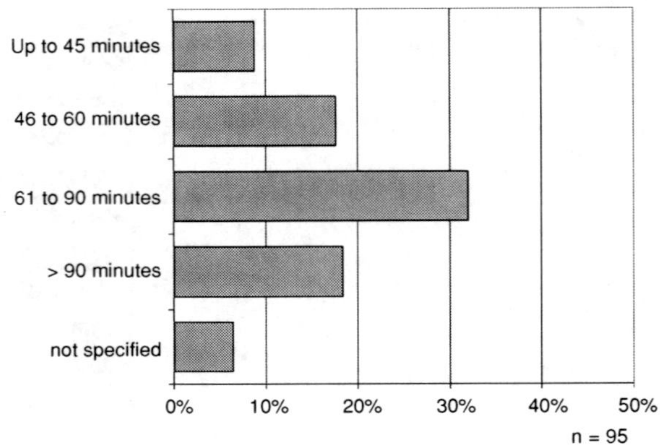

Fig. 5.29 Average travelling time to work as public transport user

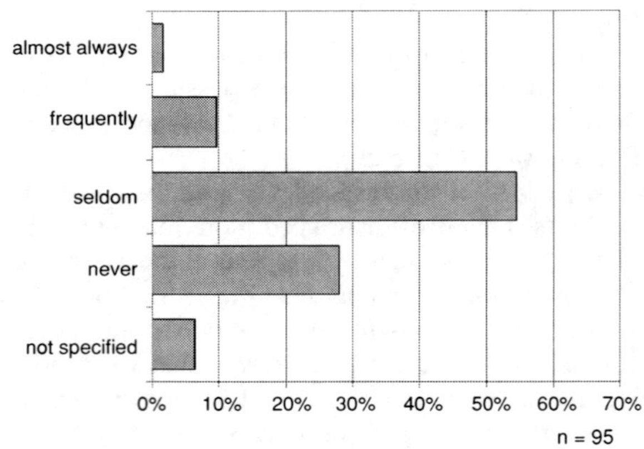

Fig. 5.30 Stopovers on the way to work or back

5.3.2.6 Open Questions

The questionnaire included some open questions to give a qualitative feedback on the CARRIVA service. Here, nine respondents or 10 % of the respondents criticize the usability of the homepage and the trip calendar at that time. Seven members motivate to expand the promotion activities to other areas than the airport. Furthermore, six respondents hope to get better information on the state-of-the-art of CARRIVA (like e.g. the current number of members). Further comments relate to the organization and offer of phone-based matching option, which has been unclear to some of the respondents. Six respondents state that they are not informed on the exact starting date of the phone-base matching service.

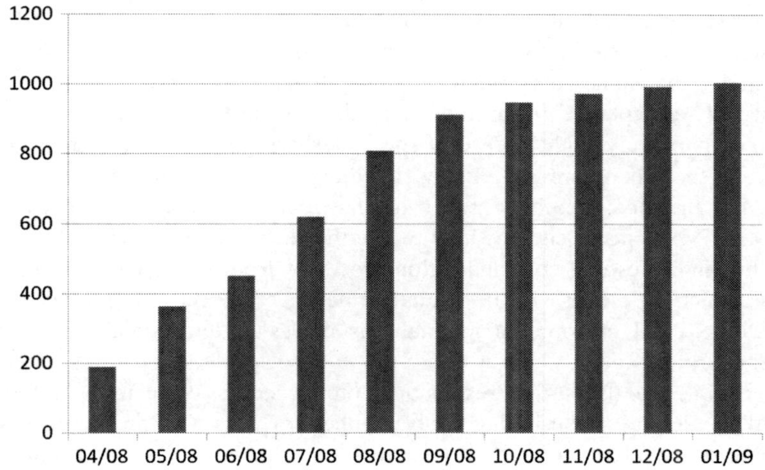

Fig. 5.31 Development of memberships

Some members indicate further reasons for registration such as curiosity or the option to be abstain from an own car on the long-term. Further respondents wish to get more information on the billing system and on the use of the trip calendar.

5.4 Conclusion from the Test Phase

During the first months of promotion activities, more than 1,000 people have registered as CARRIVA members (Fig. 5.31).

In comparison to other ridesharing activities, CARRIVA succeeded in gaining a high number of registered users, who are generally satisfied with the service. However, in spite of ongoing promotion activities the number of new registrations has leveled off after six months. About a third of all members works for Fraport and about two thirds for Lufthansa.

The surveys have revealed that employees interested in or registered for CARRIVA show a positive acceptance by attitude towards the service. On basis of the observation journals and the first survey, we can conclude that middle-aged operative staff expresses a particular interest in the Ridesharing Club. Their motivation to share a ride is primarily to share fuel costs, but environmental concerns play also an important reason to register for CARRIVA. Operative staff has predominantly reacted on the promotion activities at the airport.

With regards to the administrative staff, the picture is less clear. Most of the middle-aged executive staff has not been interested at all in ridesharing. For this target group, other promotion strategies like a premium system for ridesharing or recruitment strategies among colleagues could maybe be more effective.

In contrast, young administrative staff under 40 years seems to be particularly attracted by the CARRIVA concept. This "tech-savvy" group is predominantly interested in the phone-based ridematching system as innovation.

Not only acceptance by attitude, but also the users' acceptance by behavior is comparatively high in case of the Ridesharing Club. However, lead users point out that there are still many "teething troubles" during the launching phase. For instance, members argued did not really understand main features of the CARRIVA concept such as how to use the ridematching service. On the one hand, this indicates that communication processes from announcing an offer over the ridematching process to the actually ride have to be as simple as possible. On the other hand, information material must be easy to understand and adapted to the target group.

In general, lead users show reservations on the success or even on the functionality of the service. Ridesharing Club members express a common mistrust that has to be decisively countered in order to keep these members as customers. For gaining trust, it is essential to create a high transparency on matching options and further development step, since matching chances are indeed restricted during the launching phase. Furthermore, first "ridematching success stories" should be disseminated among the members.

With respect to the transfer of CARRIVA experiences on comparable non-profit initiatives, we can conclude that promotion and marketing activities predominantly affect acceptance by attitude, but do not solely incite the actual use of the service. In order to initiate active ridesharing among the target group, all service components from the ridematching system over communication processes to the billing system have to be consequently adapted to user demands and preferences. Particularly unique selling points like the phone-based ridematching option or cashless billing options have to be successively optimized.

Further ridesharing initiatives can also learn from CARRIVA that not a sophisticated technical system, but a high user acceptance is a necessary condition for success. This consequently includes a high usability of all technical interfaces as well as a continuous adaption to user demands. Moreover, users have to be kept at the service by a constant presence at the airport, updated news on the service and new offers like discounts for carpoolers or feedback systems. Further incentives are special parking lots or public transport tariffs for Ridesharing Club members.

In summary, conditions at locations like Frankfurt airport are particularly favorable for initiating a ridesharing service. However, ridesharing services need constant engagement and resources in order to be successful on the long term.

Chapter 6
Test Phase: Start of the Phone-Based Ridematching Service

In the beginning of 2009, the phone-based matching was about to be tested. With this test phase, the technical system should be examined concerning problems, functions and improvement possibilities on the one hand. On the other hand, the actual user behavior should be tested, because so far only the user acceptance, i.e. the general willingness to use the service, was surveyed.

The phone-based matching is a neuralgic point for a ride matching service, since a web based matching allows no feedback on the real journeys. Thus, a high number of registered members does not automatically imply that the service is frequently used. In fact, web based ridesharing service providers that have been interviewed within the project could not determine the number of actual rides initiated by their service.

In contrast, phone-based matching provides the possibility of getting a feedback for each ride, since data is available not only on offers but on the actual communication between two matching partners. This is also the precondition for billing a ridesharing contribution, which is one of the main incentives to use ridesharing services like. Indeed, the main unique feature of CARRIVA is that drivers and passengers are not required to directly exchange a contribution at each ride. Moreover, billing processes could potentially be summarized on a monthly basis. The CARRIVA concept is based on the idea to offset passengers' rides and drivers' lifts up to a limit of 20 €. This means that passenger lifts are drawn from the individual account, while giving lifts results in a surplus. If the personal account exceeds 20 €, drivers can be paid out, while passengers have to fill up their account to 20 € as soon as they have depleted their credits. Such a billing system also allows listing all lifts and routes for each ride club member on a monthly basis.

From the service provider's perspective, managing billing processes are an important means for a business model, since they allow charging a provision for each ride. From a scientific point of view, phone-based matching processes promise to provide a better database on using patterns, actual driving times, successful matching etc. than web based services, which cannot monitor actual rides.

V. Handke and H. Jonuschat, *Flexible Ridesharing*,

DOI: 10.1007/978-3-642-11345-1_6,

Summarizing, compared to web-based ridematching, phone-based ridematching has many advantages and a substantive added value not only for the users, but also for the service provider and for scientific use. However, in contrast to web-based services, phone-based ridematching is also relentless in terms of transparency, revealing data on the actual use of the service, which is assumed to be decent in the starting phase. Albeit these reservations, the project team has tested the phone-based ridematching within specific test areas with a comparatively high number of registered members. This test phase has revealed valuable information on ridematching patterns, matching processes and user behavior with regards to the use of carpooling in general and the acceptance of phone-based ridematching in particular. These results are presented in the following chapters.

6.1 Activities During the Test Phase

The test phase was prepared and carried out in February 2010. User data have been evaluated on basis of an enquiry in March 2010 as well as a Lead User Workshop in April 2010. The preparation of the test phase aimed to select areas that have a comparatively high number of members, and that are therefore suitable for testing the phone-based ridematching. In order to detect these areas, we have evaluated registration data as well as the trip calendar entries. As a result, the highest accumulation of members has been found in Moerfelden-Walldorf, Aschaffenburg, Ruesselsheim and Hofheim/Taunus.

Subsequently, the project team has called each registered member in these four areas in order to inform them on the test phase and to interview them on their riding patterns (estimated driving times, driver or passenger or both, destination at the airport etc.). Since in Aschaffenburg no one of the registered members was willing to participate as passenger in the test phase but only as driver, this area has been excluded from the test phase. From this experience, we have learned that it is crucial for ridesharing services to offer strong incentives for passengers like e.g. free memberships, or discounts for special shops etc., particularly in areas with a surplus of drivers. Summarizing, Moerfelden-Walldorf, Ruesselsheim and Hofheim/Taunus have been selected as testing areas. Figure 6.1 shows the spatial distribution and driving times of the lead user group according to their ridesharing role (drivers, passengers, alternating and members with no entries in the trip calendar):

The project team has called these lead users by phone again in order to announce the actual start of the test phase. In addition, the lead users were asked to update or confirm their driving schedule as listed in the trip calendar. Members without any trip calendar entries were furthermore asked for their reason. Figure 6.2 summarizes the fundamental reasons for missing entries in the driving agenda:

The figure illustrates the limited usability of the tip calendar on the one hand and the potential for the spontaneous mobile phone based matching on the other

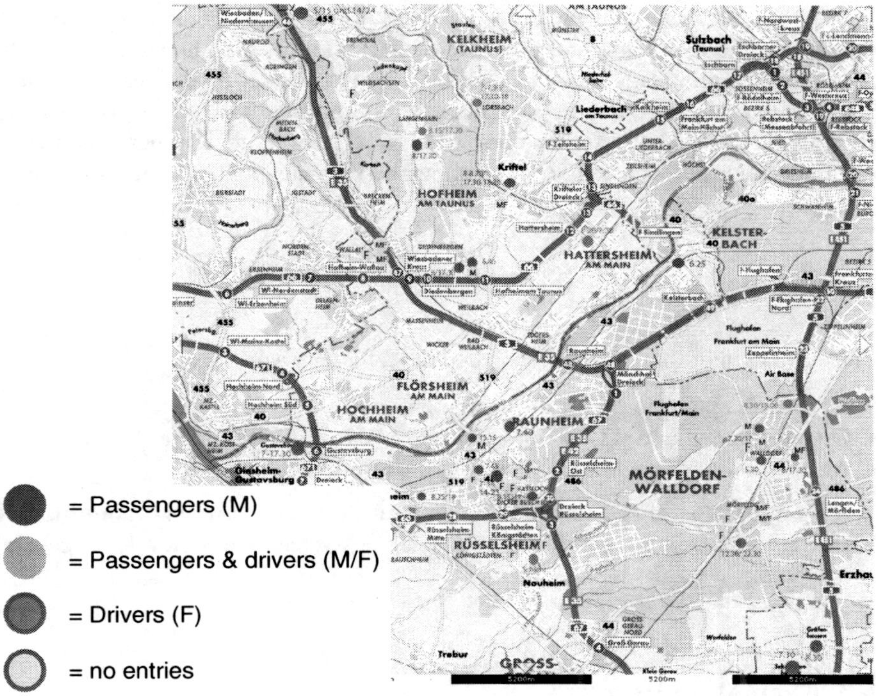

Fig. 6.1 Lead users according to ridesharing role, driving times and residence

Fig. 6.2 Reasons for missing trip calendar entries

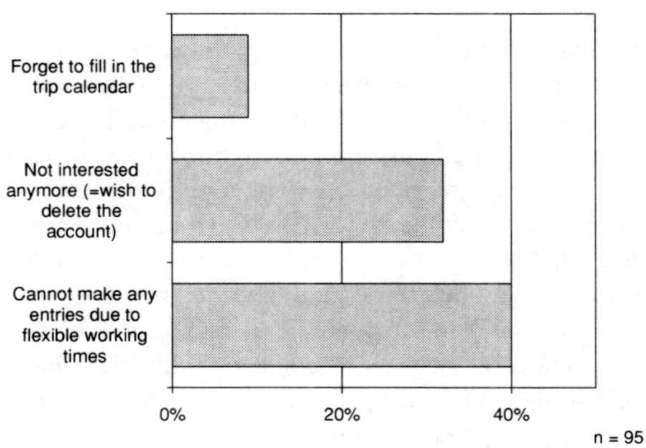

hand. It also illustrates the high proportion (about a third) of "lost" members who would like their account to be deleted. These are mostly members who have registered at an early stage to test CARRIVA, and who could not find a suitable match.

All lead users subsequently got handouts by email with all important data concerning the test phase and a detailed explanation of the phone-based matching system. Additionally, all lead users have got "personal timetables" that listed all driving times of drivers or passengers with regular driving times. This should secure a high probability for matches when calling during these times. Apart from that, members were invited to test the phone-based matching during the test phase.

6.2 Evaluation of the Test Phase

The evaluation of the test phase included the following activities:

- Interviews by phone during the testing phase,
- Organization of a lead user workshop and
- An ex-post survey in March 2009 after the test phase.

This evaluation aimed to learn more about the acceptance and motivation of lead users concerning the phone based matching system and to develop a general user typology.

6.3 Evaluation Results

6.3.1 Results of the Phone-Based Survey: Willingness for a Detour

In total, 15 passengers and 5 drivers have been interviewed. The main objective has been to enhance the matching probability and to increase entries in the trip calendar as a "backbone" of the matching system. Here, long distances between different destinations at the Frankfurt airport have turned out to be a major problem, since the interviewees were not willing to make detours to different terminals or cargo areas at the airport. A further reason for matching problems has moreover been that some entries in the trip calendar were incorrect, which led to actual mismatches, although routes and times were convergent.

Particularly the willingness to make a detour has turned out to be a key factor for a matching success. Members could determine an accepted detour as number of kilometers in their personal profile. However, this has led to the fact that most users tend to underestimate their willingness to make a detour. The majority of users have indicated that they will not accept a detour of more than 1 km, which severely restricts their chances for a match, since drivers or passengers who live more than 500 m away will not be found as a matching partner. When directly asked, if they would give a lift to a person who live in an adjacent village, most users however agreed that this would not be a problem at all, although this in some cases meant a detour of up to 10 km.

In addition, most users who accepted only a detour of 1 km stated during the interview that they would be willing to make a detour that takes much more than 5 min. Actually, the phone-based survey has revealed that only one-third of the interviewed drivers, would accept a detour of 5 min (=ca. 4–5 km), another third would accept a detour of 6–10 min (=ca. 8 km), and yet another third even more (up to 15 min and more, i.e. 12 km and more). However, the same persons stated that they would only accept a detour distance of only 2.5 km, which rather corresponds to 3–5 min driving time.

Thus, asking for the maximum duration of detour results in a much higher acceptance of detours than asking for the maximum distance in kilometers. Consequently, asking for durations instead of distances of a detour can significantly enhance the chance for a successful match. So it is recommended to generally measure the acceptance for detours as temporal dimension, since this more accurately reflects the actual willingness for detours.

With regard to the high importance of the willingness for a detour, we can also draw the conclusion that the standard setting for detours could be 6–10 min without risking acceptance problems. This considerably enhances the matching chances, since this would include all potential ridesharing partners in within a radius of about 5 km around their residence. In fact, from a psychological point of view, it is important to keep the users actively using the ridematching system and to apply to their decision, if they want to accept a riding opportunity or not. Therefore, a strategy to provoke as much matches as possible and to let the users choose between these opportunities is much more promising than a strong filter that leads to no matches at all. Thus, particularly during the launching phase, even standard settings of 10–15 min of detour are reasonable in order to incite lead user to actively use the system. In case of complaints, users will still have the possibility to adapt their preferences and choose a smaller detour.

The same applies to the indicated calling time before a planned ride, which is standardly set as 20 min. Here, we can also assume that users are much more flexible than they think in advance, if they will be "rewarded" by a matching partner. This means that users should be at least informed on all ridesharing opportunities within the next hour, even if this may be too late for them. However, a part of the users will probably rethink their choices and expand their preferences regarding travelling time and detours when having a high transparency on all possible matches.

Summarizing, the results of the phone-based survey lead to the following strategic recommendations on increasing the probability of a successful match:

- Asking for time spans for travelling instead of exact departure times in the trip calendar,
- Asking for the duration instead of the distance of a detour for picking up passengers,
- Clustering adjacent destinations at the airport,
- Initiating to use the phone based ridematching particularly during rush hours,
- Developing personal timetables with regular rides.

6.3.2 *Results of the Lead User Workshop*

The lead user workshop was held at the Frankfurt airport in May 2009 after testing the phone based ridematching system for two months. Within this workshop, seven lead users have expressed their experiences with the matching system. The statements were discussed and the lead users have developed recommendations from the user's perspective for improving the matching system. Apart from experiences with the matching system, the lead users were also asked to comment the promotion activities at the airport, the CARRIVA homepage and CARRIVA's communication activities. Within the following sections, we have summarize the main outcomes from the lead user workshop.

6.3.2.1 Motivations to Use CARRIVA

The lead users gave the following reasons for participating in the CARRIVA Ridesharing Club:

- *Environmental protection associated with a higher vehicle occupation rate*: The low grade of occupation was conceived as wasteful and ecologically unsound behavior.
- *Avoiding traffic jams*: In spite of a permanent extension of traffic infrastructures, essential routes to the airport are particularly prone to traffic jams. Ridesharing can potentially contribute to avoid traffic jams and reduce travel times.
- *Corporate identity reasons*: One interviewee has been working for the environmental management section of his company and stated that carpooling schemes can positively contribute to the corporate image.
- *Travel cost reduction*: Reducing travel costs is a substantial reason for participating in the CARRIVA Ridesharing Club. Interviewees have estimated saving of up to 300 € per month.
- *Alternative to public transport*: Particularly users with a deficient access to public transport hope for a more convenient travel option by ridesharing.
- *Deficient transport options at the airport*: Ridesharing has been regarded as an alternative to inconvenient airport shuttle system.

6.3.2.2 Promotion Activities

When asked for the first contact with the CARRIVA Ridesharing Club, the lead users named the following information sources:

- Lufthansa and Fraport have announced the CARRIVA offer on their intranet starting pages for several days.
- The promotion activities of CARRIVA have encouraged most of the interviewees to register for the ridesharing service.

- Furthermore, workmates have recommended the service.
- The environmental management section of some companies has also informed on the CARRIVA offer and appealed to take part in the ridesharing club.

The lead users have particularly discussed a closer cooperation with companies located at the airport, since information via the employers intranet seemed to be particularly effective and trustworthy.

Employers have a direct access to staff data like the home address, and could therefore directly contact employees in a certain area. They can moreover easily start a short survey, and ask for the willingness to participate in a ridesharing scheme. In this context, data security has been an issue, which has been discussed controversially. Furthermore, it has been recommended to contact smaller employers like e.g. shipping companies as well. Further cooperation options have been seen with public transport companies, who could actively promote ridesharing as part of intermodal travelling. In this context, promotion activities at train stations and bus stops were considered as a useful activity to particularly address public transport users for the Ridesharing Club.

6.3.2.3 The CARRIVA Homepage

When the lead user workshop took place, the CARRIVA homepage has been a first release for the pilot project phase. Accordingly, the lead users have criticized several problems and bugs like a complicated navigation or an "unprofessional" appearance. In general, the workshop participants have recommended to focus on the essentials and to avoid too much extra information. Though, the lead users have classified costs for rides, information on members living nearby and hints on matching chances as most important information that should be available and easy to find on the homepage. In this context, the lead users particularly appreciated the visualization of registered members on the map, but also recommended to add travel times, if available.

Summarizing, the discussants particularly demanded a high transparency on riding opportunities as reference on matching chances at certain locations and times. These recommendations have been reported to the developers who considered them in the redesign of the homepage.

6.3.2.4 Registration Process

The lead users have not experienced any general problems with the registration process. The declaration of preferences has been complete and nothing was missing. Still, the lead users had some concerns regarding the registration process:

- The required data like telephone number, working times etc. has been estimated sensible data and a high data security has been given a high priority.

- The detailed request of preferences has been a little bit confusing and the relevance for the matching process has been unclear. Moreover there has been the concern that too many preferences could impair the chances for a matching. One discussant has summarized that one should "rather be matched with someone who lives 12 km away—you can still always cancel this option!"
- The option to state the willingness for detours has been criticized as well: "A detour in kilometers is not equivalent to the duration of a detour!" Moreover, "10 km can be acceptable in rural areas, whereas in Frankfurt, this would not be acceptable." Furthermore, the willingness for a detour at the airport has been considered lower than at the place of residence. While at the place of residence the acceptance of a detour of 10–15 min is acceptable, a detour of few kilometers at the airport has not been tolerated.
- Some participants positively pointed out the additional registration opportunity via the CARRIVA hotline, since the CARRIVA staff could answer special questions, which has been regarded as helpful and assuring.

In order to improve the registration system, the workshop participants have particularly recommended to abbreviate the preference statements as well as to separately ask for the willingness to make detour for each ride. Furthermore, it has been recommended to set detour standard at 15 min at the drivers' home address. Passengers however do not need to be asked for their willingness to make a detour, since they will probably not be willing to make a detour of more than 500 m. The same applies to destinations at the airport: the system should assume an acceptable deviation of 500 m by default from the indicated airport destination. Furthermore, drivers should be asked for their willingness to pass by adjacent destinations at the airport. Here, the system needs more information on main routes serving different airport locations.

6.3.2.5 Trip Calendar and Personal Timetable

Filling in the trip calendar has not caused major problems. The lead users have particularly emphasied the personal timetable as an appropriate instrument to increase the transparency on mathcing options. However, for employees with flexible work times, it is too complicated to use trip calendar, since this requires to make three subsequent entries per hour. The pop-up windows informing on successful matches have been perceived as disturbing. They particularly impair the motivation to use the service, when they only inform that no matching partners have been found. Though, the pop-up-information has been discussed controversially, since an appropriate information on mismatches like e.g. members on the same route, but at another time or potential matches that have been excludede due to personal preferences (e.g. a matching driver who likes to smoke) have been considered as very important.

For optimizing the usability of the trip calendar, the lead users have pointed out to adapt it to employees with flexible working times by finding a possibility to enter time spans. Here, the discussants have proposed to integrate an option for announcing flexible trips in general. It has also been recommended to design the selection opportunities as a dialogue, e.g. asking for the work model (flexible, office hours etc.) and then lead to the relevant page to fill in the trip calendar.

Furthermore, it has been recommended to deactivate the pop-ups and rather improve the transparency on potential, but yet not matching partners, e.g. by giving a link to a personal timetable on the required route. This would give the users the opportunity to question and adapt their travelling preferences in order to enhance the chance to find a matching partner.

6.3.2.6 Matching Process

With regard to their experiences with the matching processes, the workshop participants have named the following aspects:

- *Incorrect matchings*: One user claimed to have received matching proposals that haver matched neither spatially nor temporally. Apparently, there has been a mistake in the personal settings, which have not been adequately explained to the user. A further user stated to have only received information on return trips but no trips to the airport.
- *Regular e-mails*: Some users have daily received an e-mail with regular trips on their route, which has been perceived as annoying and made the user delete his trip calendar entries completely. Here, users should have the opportunity to cancel the e-mail option, if e.g. the according matching partners do not answer on requests to share the ride.
- *Meeting points*: Arranging a meeting as well as the lift itself has been described as conveniently and without any problems.
- *Costs*: The proposed share for passengers have been estimated as rather high. However, during the test phase, the lead users have shown a high tolerance: "It is always give and take—sometimes I am the driver and get paid, sometimes I′m the passenger and pay."

In order to improve the matching process, the participants have generally recommended to make the processes as transparent as possible and to "bring the partners together" instead of filtering matching opportunities beforehand. Hence, it is a priority to communicate potential matches, even if they are not optimal. The tolerance concerning incorrect matches seems to be quite high during in the initial phase, while no matches are in contrast regarded as "frustrating" experience. The lead users have expressed a general mistrust in the technical system, while direct contact to other members has been described as convenient experience. In general, matching chances should be communicated transparently and user-friendly: "The information has to be eye-catching and you shouldn't have to look for it for a long time!" With regards to the daily e-mails, the workshop participants have recommended to change from a nocturnal matching to a weekly or monthly check, in order to reduce the number of automatically generated information emails.

6.3.2.7 Expansion of the Phone-Based Ridematching

Optimizing and expanding the mobile phone based matching has been regarded as important step for the acquisition of new members. The phone-based matching is a unique selling point and demonstrates that the CARRIVA concept could really initiate flexible ridesharing. Only the expansion of the mobile phone based matching could reach users, who want to spontaneously share a ride. All workshop participants have agreed that it is more reasonable to offer the service and make also missing matching possibilities transparent, than "retaining" the mobile phone based matching and wait for reaching a theoretical critical mass. The expansion of the mobile phone based matching service should be accompanied by further promotions activities, which inform on the new opportunities for flexible ridesharing and communicate that CARRIVA is constantly optimizing its offer. This could also help to interest new members in the service. Unsuccessful calls are not regarded as a problem, because the current members are aware of initial problems during the launching phase. However, "the best publicity is still a successful service".

6.3.2.8 Ranking of the Lead Users' Recommendations

With regard to the abundance of proposals and long dicussions, the workshop participants have been asked to prioritize their demands. Consequently, the following issues have been given a high priority:

1. Nearly all participants considered a transparent matching process as the most important recommendation. Here, the accuracy of matching results should be communicated differentiated and also well-nigh matches should be communicated. This should enable the user to rethink their preferences and to adapt them temporally or spatially.
2. Half of the participants considered it the most important strategic approach to rather bring members into contact for negotiating rideshare opportunities than excluding potential matching opportunities in advance.
3. Most participants also claimed that the spontaneous ridematching option via mobile phone should be introduced to all regions as soon as possible.
4. Further recommedations included:

   - Expanding the cooperation with local companies and employers,
   - Improving the design and usability of the homepage,
   - Changing the automatically generated information on ridesharing opportunities from daily to weekly or monthly,
   - Reducing the recommended share for passengers.

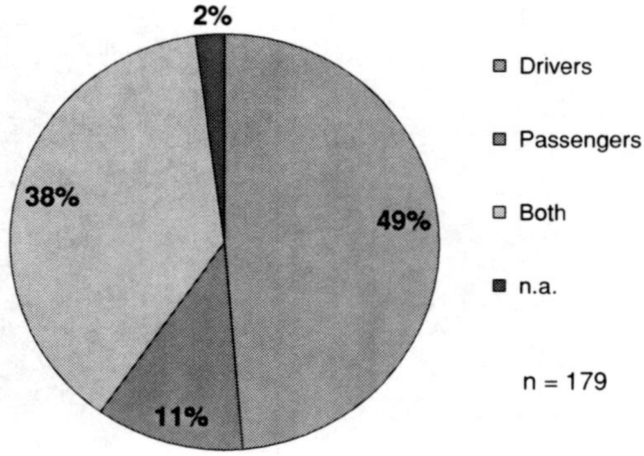

Fig. 6.3 Proportion of drivers and passengers

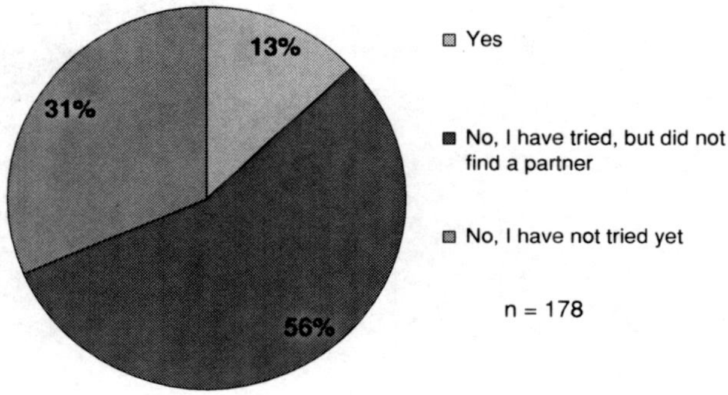

Fig. 6.4 Use of the phone-based ridematching option

6.3.3 Results of the Ex-post Survey

After the test phase, we have started an online survey on the using patterns among all members of the Ridesharing Club., 179 users of a total of 1,000 members have participated, which corresponds to about 20 %. The survey aimed to give insight on the distribution of user types, on the main ridematching patterns as well as on the trips matched by CARRIVA (Fig. 6.3).

The proportion of user types shows that nearly half of all respondents (49 %) are exclusively drivers. The second largest group (38 %) consists of users that are registered as both, drivers and passengers. Exclusive passengers have only very share of 11 %. In total, only 2 % of the users have not made any statements concerning their using type (Fig. 6.4).

Regarding the phone-based matching, it is evident that only half of the members have tested the phone based matching option. This confirms that the users need more incentives to actually use the phone-based ridematching. Though, only a small share (13 %) of members have succeeded in finding a matching partner, which indicates that there is still only a small chance to get a lift via the phone-based matching. In contrast, it also means that nearly a quarter has actually shared

Fig. 6.5 Use of ridematching options

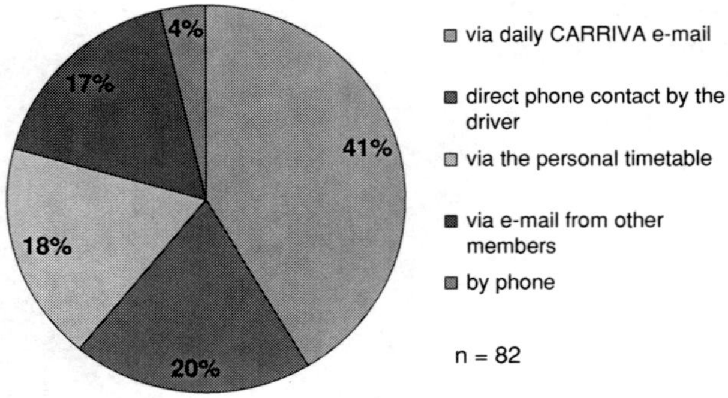

Fig. 6.6 Number of trips arranged by CARRIVA

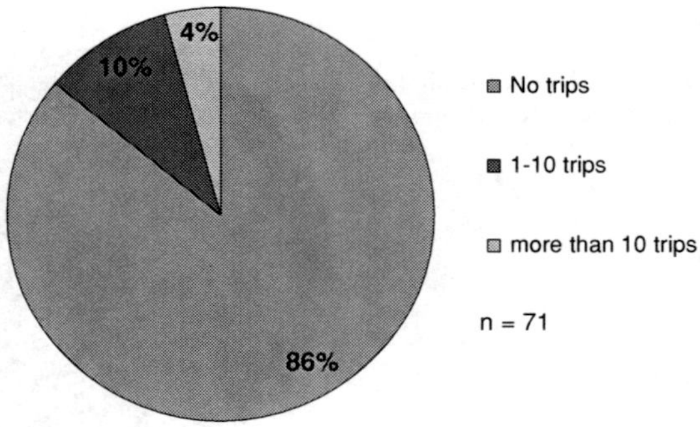

a ride via CARRIVA. One reason for the small share of actual rides could be that the matching system has not been optimal at that time on the one hand. On the other hand, a ratio of 1:4 could also refer to general chances to get a match, regardless if the user has used the phone-based or the web-based ridematching option (Fig. 6.5).

Most of the users (41 %) have contacted other members by email on basis of the daily email function. Furthermore, 18 % of all respondents have used the personal timetable to contact potential ridesharing partners by email, although this ridematching option has only been available after the start of the test phase. Therefore, three-thirds of all matchings have been arranged by email after getting information on potential partners either by the daily CARRIVA e-mail, the personal timetable or a direct e-mail request of a member. This emphasizes the importance of both push information as an incentive to use the ridesharing service and e-mail communication.

Though, in spite of the restricted usability, 4 % of all respondents have still used the phone based ridematching option. Further 17 % of the respondents have directly contacted passengers by phone after receiving information on ridesharing requests either by the personal timetable or the daily CARRIVA information e-mail. Consequently, nearly a quarter of all members have arranged their trips by phone instead of email.

In total, the number of actual trips is still very low (Fig. 6.6).

More than 80 % of all respondents have tried to find a trip, but have not actually shared a ride yet. Still, with respect to the early stage, it is remarkable that nearly one out of six respondents have already shared a ride due to CARRIVA and already 4 % have been shared more than 10 rides with CARRIVA members.

6.4 Ridesharing User Types

There are five different types of ridesharers: The frequent driver or passenger, the spontaneous driver or passenger as well as the flexible driver. In this context, one has to take into consideration that the user typology does not correspond to individual users: While a person can be a frequent driver or passenger in the morning he or she could change to a spontaneous driver or passenger on the ride home.

6.4.1 The Regular Driver

The regular driver uses to tale his personal car always on the same route at fix times. He shows a weak willingness to take a lift and abstain from using the private car. Regular drivers can be frequent ridesharers as long as giving a lift does not restrict their travelling routines. For a regular driver, the trip calendar is perfect for finding passengers. The reduction of costs is an important incentive for regular drivers to share a ride.

6.4.2 The Regular Passenger

The regular passenger does not use or even own a private car. He predominantly uses public transport on the same route at fix times each day. Ridesharing is only of minor interest for regular passengers, because they usually can fulfill their mobility needs with public transport. However, there are still potentials to interest regular passengers as ridesharers, if a shared ride has advantages in comparison to the public transport, e.g. by being more comfortable or cheaper. Regular passengers will probably rather use ridesharing services occasionally in special situations like public transport problems or at bad weather conditions.

6.4.3 The Spontaneous Driver

The spontaneous driver owns a personal car. He is not willing to fundamentally change his mobility behavior, but will be open to offer lifts for other members spontaneously. Spontaneous drivers are an important target group for phone-based matching, since web-based matching needs too much planning in advance for the spontaneous driver. Spontaneous drivers will particularly appreciate the free phone based matching option by short code calls as well as an indirect billing system via the member account.

6.4.4 The Spontaneous Passenger

The spontaneous passenger generally uses public transport, but may also possess a private car. He is willing to use ridesharing services as a passenger, but needs to be flexible with regards to the travelling times. Spontaneous passengers need strong incentives to use ridesharing services, such as a reduction of costs as well as in a reliable system. This user type is assumingly a "laggard", who will register rather late as Ridesharing Club member. In fact, our surveys have shown that most passengers of the test phase have not been very flexible with regards to their travelling time, and expressed that they need a security to get to work in time.

6.4.5 The Flexible Ridesharer

The flexible ridesharer is the ideal user of flexible ridesharing concepts like CARRIVA, since they are not only flexible in terms of travelling times, but also with regard to their role as driver or passenger. Flexible ridesharers determine spontaneously before departure to work to drive their car or take a lift. They are particularly interested in a phone based ridematching option, since this promises the highest flexibility for spontaneously arranging a lift. From the provider's perspective, this user type is the most interesting target group for CARRIVA, since flexible ridesharers are likely to change their mobility behavior from single occupation car use to ridesharing—if they can use an appropriate flexible ridesharing service. In addition, this commuter group constantly grows due to a rising importance of flexible working hours, but also flexible working places (at home, mobile work etc.)

Chapter 7
Recommendations for the Dissemination Phase

In order to outline the basic development steps to successfully introduce a mobile phone based ridematching service, the following chapters emphasize significant aspects and requirements along substantial implementation phases. Principally, fulfilling main requirements depends primarily on the basic conditions at the place of implementation and the local organisational structures and processes. As a general rule, it should be considered that the implementation ought to be carried out successively, taking account of given mobility patterns as well as of the acceptance and motivational structures of the relevant target group.

7.1 Step 1: Optimising the Technical System

It can be assumed that the continuous digitalisation and the increasing distribution of information and communication technology (ICT) will lead to rising technical possibilities on the one hand, and ICT will gradually become a natural constituent in everyday life of huge parts of the population on the other hand. Powerful mobile devices become increasingly available. This results in new possibilities of using specific software applications for a mobile phone supported agency service, which allows an integration of the agency service in the other digital services for final customer use. Concerning the technical acceptance, one can assume that for particular routine activities like registration, announcement of journeys, using abbreviated dialling numbers for special destinations etc., users will gain more experience regarding the handling of the devices and applications. However, enhanced experiences of customers with digital services and smart phones will also raise e expectations concerning comprehensibility, clarity, usability, design etc. These expectations are in the same way requirements for the establishment of the technical system at the implementation of a mobile or Smartphone supported ridematching service.

V. Handke and H. Jonuschat, *Flexible Ridesharing*,
DOI: 10.1007/978-3-642-11345-1_7, © Springer-Verlag Berlin Heidelberg 2013

7.2 Step 2: Continually Optimising the Service Concept

In order to successfully develop the service concept for the implementation of a phone based ridematching service, users explicitly need an adequate support and involvement of service staff at the location of implementation. This can vary considerably according to the existing organisational structure. Therefore, a preferably detailed and exact analysis of the organisation is needed, not only with regards to the structure and the processes, but also with respect to existing mobility patterns. Here, main employers and local companies are important partners for addressing commuters. This is not only a main instrument to ensure the support of promotion activities, but also to underline that the ridematching service is trustworthy.

The service concept must be designed as an adaptable system. In a first development step, it is important to analyse local framework conditions like mobility patterns and employment structures. Following activities will focus on addressing motivations and needs of the target groups and their practical experiences with ICT or ridesharing. Finally, the service concept has to be continually corrected, developed, improved and expanded. Basically, it is recommended to develop the service concept with the help of a business model as a basic core offer that however, allows as much further adaptation options as possible. An essential part of the service concept is to decide on the institutional framework (association, company etc.). Finally, cooperation with local organisations and companies has to be clear and possibly supported by a contract. In general, it is important to rely on a progressive process, which means to start with a core service concept at an early stage, and then optimize it successively according to the using experiences and feedbacks. Incentives like special parking lots for ridesharers are also part of the service concept.

7.3 Step 3: Promotion Activities for Acquiring Lead Users

By addressing first lead users, service concept is put to the test, because users will find easily find any deficiencies. It is therefore essential to directly react on the user feedback, since otherwise there is the danger of losing impatient lead users as well as impairing their willingness to use the ridesharing service in general.

Before starting canvassing, a multitude of activities and tasks have to be planned: Besides producing information material and the service's homepage, promotion activities include a broad information campaign addressing all local commuters. Using established communication paths is an important link to the future customers. Therefore, communication on the new ridesharing offer via internal mailing lists or intranet communication, but also articles in company journals are important and trustable ways to reach the target group. It should be the goal to reach as much commuters as possible and to inform them on the existence of the ridesharing service. These initial activities should be accompanied by a user-friendly homepage, simple contact tools and competent contact persons to provide personal help in case of further requests.

The further acquisition of members should be reflected by a strong presence on site. Depending on local structures, different strategies can help to gain attention for the ridesharing service in order to reach as many commuters as possible. Here, a non-profit concept has clear advantages to commercial concepts, since the psychological threshold for being a "club member" is much lower than committing to commercial offers.

The first contact to new members has also to meet different requirements. Appropriate measures are for example short encounters at parking places or entries in terms of e.g. handing out information material. Detailed information can be given in places where people do not pass-by, but stay for a while, like canteens or at bus stations. Company events represent also an important opportunity for reaching the target group. Depending on the nature of the event, like e.g. internal meetings, company celebrations or sports' activities, potential users often have the time to get information material and to ask questions. Principally, information events concentrating on the ridesharing service itself promise to be particularly effective. Yet, this demands a thematically competent preparation and a high readiness for support by the company management, especially when working time is lost. At events, the system can be demonstrated, simulated, visualizes, and tested by the users.

Finally, promotion activities can be organised as a dynamic system that is powered by the users. In this context, social networks and intranet communication can be actively used to initiate promotion activities "from member to member".

7.4 Step 4: Integrating Motivations and Needs of the Target Group

Basic framework conditions for motivation and needs of the target group are already part of the organisational analysis while developing the service concept. Further details on times and routes of commuter patterns, and especially on user-specific motivation and needs can be gathered on site as part of the promotion activities. Besides, lead users constitute an important information source concerning the detailed analysis of needs and demands, drivers and obstacles. This information is an essential framework condition for the introduction, adaptation and stabilization of the ridesharing service.

7.5 Step 5: Testing the Service with Lead Users

For a quick start of the ridesharing service, lead user studies play an important role. They can help to understand local concentrations of members, preferred routes, and time preferences of journey offers. Therefore it is essential to continuously evaluate the dissemination of the service. Information on lead user preferences can also contribute to define a suitable testing area or selected routes with a sufficient chance of getting a matching ridesharing partner. In order to optimise the ridematching service and prepare a complete roll-out, lead users should be surveyed with regards to their opinions both on the service features and on strategies for further promotion strategies.

During the test phase, the technical system and the service concept should be continuously improved and expanded according to the evaluation results. Here, one should consider that a phone based ridematching service demands completely new user behaviour, and lead users cannot build upon former experiences. Furthermore, mobility patterns differ considerably from one region to another, and user acceptance is also dependent on each service concept. Therefore, only a "real life" test can bring adequate results on actual acceptance both by attitude and by behaviour.

In order to avoid annoying present service customers, there has to be a permanent quality control and management during the test phase, including a systematic feedback system for the lead users on how their experiences have been used to adapt and optimize the service and the technical system. The result of the test phase is a customer-friendly service environment and a fail-safe technical system.

7.6 Step 6: Expanding the Service Offer

The optimized ridematching service is now ready to be intensively promoted in order to reach as many customers as possible. For expanding the service offer, three aspects are essential:

- The promotion activities should be adapted to each customer group. This implies to develop different strategies e.g. for passengers, women, young professionals, tech-savvy groups etc. Furthermore, promotion activities should also concentrate on certain locations (e.g. bus stops in certain areas with many members or party locations) with a high chance of reaching the relevant target group.
- Acquiring new members by existent members is a very efficient way to expand a service offer. Here, incentives for acquiring new members like premiums or shop discounts can help to activate existing ridesharers who indeed profit from a broader offer of lifts.
- Further incentives like free or attractive parking lots will also help to increase the number of ridesharers and thus, the number of offers. During the expansion phase, public and commercial partners will be more open for negotiations on such incentives.

The strategic development of a broad incentive system is an essential success factor both for acquiring new members and for customer relationships in general. Especially on the long term, attractive incentive systems are promising instruments to constantly get new offers for ridesharing trips. These incentive systems however, depend strongly on the location and potential local partners to contact. Since local authorities have an interest in avoiding traffic congestion and promoting alternative transport options, they could be a powerful partner for the ridesharing service. Furthermore, partners from the "green mobility" sector like carsharing companies or IT service providers who develop mobility applications could also be interested in cooperation.

7.7 Step 7: Establishing the Billing System

Cashless payment of shared rides is one of the most powerful incentives for ridesharing services. Thus, implementing the billing system is an important step to expand the ridematching service. Here, the mobile phone-based ridematching process allows to exactly monitor actual trips, and consequently to calculate the exact fare for the given route. In this context "member accounts" can be an effective incentive for drivers to also take a lift now and then, since giving a lift can be balanced by taking several lifts. Only the account has a certain surplus or credit, e.g. 20 €, passengers have to pay their lifts or drivers will get their fares.

For the service provider, though, establishing a billing system represents the basis for a business model, since they could levy a certain commission on each transaction. In contrast, from the user's perspective the accounting system is also a sensitive issue, since cashless payment affects highly sensitive and vulnerable information like bank details or account data. For this reason, data security and a fail-safe billing system is a precondition for establishing cashless payment systems.

7.8 Step 8: Constant Monitoring and Quality Control

Eventually, the service has to be constantly monitored in order to keep its high quality. In order to keep the ridesharers active, new service features like connections to social networks, ridesharing Smartphone applications or GPS-based features (e.g. for detecting Ridesharing Club members nearby) will make ridesharing more attractive. Although this means a constant investment in the service, an expansion of the service to other areas involve less costs and resources than the initial development of technical components and the initial service concept.

Chapter 8
Assessment of Sustainability Impacts

The environmental impacts, which could be avoided by the implementation of the ride sharing service at the Frankfurt airport, can be quantified by methods of life cycle analysis. To carry out this analysis, some assumption and a modelling have to be made in a first step. In general, it is assumed that about 70,000 people are working at the Frankfurt airport, of which around 75 % go to work by their own passenger car. Employees using duty vehicles, motorcycles, mopeds or bicycles remain unattended. With 1.4 passengers per vehicle (BMVBW 2005), that means daily commuter traffic of 37,500 passenger cars. To estimate the environmental benefit achieved by the ride sharing service, it is assumed that the number of the passengers per vehicle rises up to four persons per car. That means that the daily avoidable amount of vehicles in using has a determinate size of 24.375 cars. For this amount of avoided cars an average daily driving performance of 50 km is assumed. Concerning the used fuels, it is assumed that 33 % is diesel and 67 % is petrol. This splitting relation equates to the national fuel mix in the year 2005 concerning the activity rates of the last validated national energy balance (AGEB 2006) of the mobility sector in the central emissions inventory of the national system of greenhouse gas emission (see Jörß et al. 2010). The share of bio diesel is not specifically considered because of its small share of 2 % of all fuels used by passenger cars. This share is added to the conventional diesel share. Fluid gas is also not considered separately because of its small share of about 0.01 % (Jörß and Handke 2006).

The modelling is adequate to a maximal feasible reduction case. The assumed number of four passengers per vehicle is close to the maximal car use by ridesharing and is therefore very optimistically. Taking into account these assumptions, a ride sharing agency service at the Frankfurt Airport could save about 233 TJ Diesel und 502 TJ per year. That means an overall saving of 0.1 % of the national fuel consumption in the year 2005 and a saved driving performance of 266 Mio. km.

According to the reduction technology of the vehicle, an average reduction technique is assumed. Vehicles without reduction technologies are not considered because of their small share of 6 % of all passenger cars in the year 2005. Furthermore, this share will decrease because of the incremental technology

V. Handke and H. Jonuschat, *Flexible Ridesharing*,

DOI: 10.1007/978-3-642-11345-1_8,

Table 8.1 Through the agency service avoidable air pollutants

CO_2	NMVOC	NO_x	NH_3	SO_2	TSP	PM $_{2,5}$	PM $_{10}$
54.295 t	22 t	66 t	5 t	0.3 t	19 t	5 t	19 t

Source Own calculation based on the central system of emission inventory (ZSE) 2005–2010 u.a

development of the automotive sector and the national vehicle-environmental-bonus, which subsidises the purchasing of a new and more environmental friendly car. As polluter groups the differentiations of the TREMOD system are used and the specific share of the operation modes highway with 32 % as well as urban and extra urban traffic with 34 % each are assumed.

The used emission factors distinguish between vehicle type, fuel, operation mode and of course pollutant, concerning the numerical values used by the national system of emissions inventory for the year 2010. In the case of the pollutant CO_2 only the type of fuel is used as differentiation. In case of particular matter differentiated in the sup types TSP, $PM_{2,5}$ and PM_{10}, the exhaust emissions are supplemented by the friction emissions from the road surface, the tyres and the brakes, as well as emissions caused by resuspension (Handke et al. 2007). In case of the NMVOC-emissions the evaporation of petrol by fuel-filling are not considered (Handke and Jörß 2007). Taking into account the above-mentioned assumptions, the following table shows the yearly amount of emissions that can be avoided by using the agency service (Table 8.1).

The numerical values in the table above underline that there are quantitatively relevant potentials on avoidable air pollutants, which could open up by the use of an agency service at the Frankfurt airport.

The order of magnitudes of avoidable air pollutants follows typical shares of the traffic sector on these pollutants (see Handke et al. 2007). The highest amount of avoidable pollutants is reachable for CO_2 followed by NO_x and NMVOC. The avoidable amount of SO_2 and NH_3 is rather marginal. The avoidable amount of particular matters is moderate from a national point of view. Nevertheless, from a regional point of view, the reduction could have a relevant impact on the immission load.

To be able to quantify avoidable environmental pollution by CARRIVA beyond the classical air pollutants, the method of life cycle analysis was applied. On the basis of the 266 million kilometers travelled, which can be saved by using ride-sharing at the Frankfurt airport, the accordingly avoided environmental impacts can be calculated. Therefore, the LCA-database ECOINVENT, dataset V2.2 (2010), current database query July 2010, was made use of. Applied were the inventory data of the operation of passenger cars with reference to the European average fleet value in the year 2005. The kilometers driven per vehicle were used as functional unit. The inventory data comprise the environmental impacts caused by fuel consumption. Considered are the direct gaseous emissions caused by combustion as well as particulate matter and heavy metal emissions. Concerning the particulate matter emissions, the emissions from abrasion are also considered. The construction and provision of traffic infrastructure are excluded.

Table 8.2 Environmental impacts to avoid by ridesharing

Effective category	Value	Unit	Source and evaluation method
Climate Change	62	kt CO_2-Eq	IPCC 2007 GWP 100a
Cumulative energy consumption (fossil, nuclear and renewable)	819	TJ-Eq	VDI-guideline 4600
Eutrophication-potential	28	t PO_4-Eq	CML 2001-
Eutrophication of waterbodies	7	t PO_4-Eq	IMPACT 2002+ (Midpoint)
Acidification of waterbodies	168	t SO_2-Eq	IMPACT 2002+ (Midpoint)
Reduction of abiotic resources	391	t Sb-Eq	CML 2001
Photochemical ozone formation	47	t Ethylene-Eq	EDIP
Stratospheric ozone reduction	8	kg CFC-11-Eq	EDIP, ODP 20a high NOx POCP
Human health	965.405	eco-Points	eco-indicator 99, (H,A)
Human toxicity	12.625	t 1,4-DCB-Eq	CML 2001 HTP 100a

Table 8.3 Avoidable consumption of resources per year due to ridesharing

Consumption of metallic and energetic resources	Total	Dimension	Source
Consumption of metallic resources			
Aluminium	1.368	kg	EDIP
Cadmium	1.37	kg	EDIP
Cobalt	0.15	kg	EDIP
Copper	782	kg	EDIP
Gold	0.03	kg	EDIP
Iron	143.491	kg	EDIP
Plumb	120	kg	EDIP
Manganese	212	kg	EDIP
Quicksilver	0.02	kg	EDIP
Molybdenum	24	kg	EDIP
Nickel	2.486	kg	EDIP
Palladium	1.04	kg	EDIP
Platinum	0.03	kg	EDIP
Silver	0.07	kg	EDIP
Tantalum	0.03	kg	EDIP
Tin	1.38	kg	EDIP
Zinc	521	kg	EDIP
Consumption of energetic resources			
Brown coal	512	t	EDIP
Coal	403	t	EDIP
Natural gas	1.126	t	EDIP
Petroleum	17.685	t	EDIP

The evaluation of the environmental impacts is carried out according to appropriate evaluation methods. The following table provides an overview of selected evaluated environmental impacts, which can be avoided in one year by using the switching service at the Frankfurt airport (ECOINVENT dataset V2.2 2010, own calculations) (Table 8.2):

To also account for the shortage of resources—recently characterized as especially relevant—the consumption of some selected metallic and energetic resources was calculated as well. The following table accordingly provides an overview of the results (ECOINVENT dataset V2.2 2010, own calculations) (Table 8.3):

This table impressively marks the resource relevance of the motorized individual transport. The largest amount of consumed resources relate to iron, aluminium and nickel, followed by the metals Copper and Manganese as well as the heavy metals Plumb and Zinc. Astonishing are the still appreciable amounts of both pollutants like Cadmium und Quicksilver and of recyclable fractions like Gold, Silver, Platinum and also Tantalum. The table also strikingly underlines the massive consumption of fossil energy resources.

Chapter 9
Outlook

Carpooling can be an important element of a sustainable mobility system of the future. With an increasing number of passengers per vehicle, traffic can be reduced resulting in environmental benefits. Technical improvements and the growing availability of mobile internet, as well as increasing prices for fossil fuels, ridesharing will certainly gain further attention in future. With regards to the trend of increasingly flexible working patterns, only flexible ridesharing services using mobile devices for finding matching partners can fulfil the users' needs for individual mobility.

The technology for phone-based ridematching is already state-of-the-art and will continuously be improved and completed by smartphone applications. Significantly more difficult is the successful implementation of framing service concepts. Here, open concepts combined with successive monitoring and optimization can help to react on actual mobility needs as well as on motivations of the users.

However, commuting patterns do not only consist in transportation from A to B. Therefore, it is important to actively integrate users in the further development of the system by evaluating their demands and their preferences on the one hand and by implementing constant feedback systems. Furthermore, ridesharing services have to be adapted as far as possible to local structures and frameworks, depending on different spatial and socio-economic factors like main routes, central employers or public transport systems.

The analysis of commuting patterns at the Frankfurt airport has shown that mobility behaviour is adaptable and subject to negotiation processes to a great extent. Neither the status as driver or passenger nor departure times or routes are fix, but users are indeed willing to adapt their behaviour if this enhances the chance for a match.

Thus, analysing mobility patterns and demands of commuters as well as a constant quality control is the main success factor for ridesharing services. Only by taking the customer needs seriously, flexible ridesharing will have a future.

V. Handke and H. Jonuschat, *Flexible Ridesharing*,
DOI: 10.1007/978-3-642-11345-1_9, © Springer-Verlag Berlin Heidelberg 2013

References

Arbeitsgemeinschaft Energiebilanz (AGEB) Vorläufige Auswertungstabellen zur Energiebilanz von 1990 bis 2005 Table 2.1: Struktur Energieverbrauch. 09/2006 der Arbeitsgemeinschaften Energiebilanzen e.V., Berlin (2006)

Wegmann, A., et al.: Ridesharing in den USA, Kanada, Neuseeland und Europa bis 2007 Erfahrungen, Projekte, Erfolgsfaktoren—Literatur- und Internetanalyse im Rahmen der Machbarkeitsstudie RMS vom August 2007. Eigenverlag der Sozialforschungsstelle der Universität Zürich. 2007

Artho, J., Haefeli, U., Matti, D.: Evaluation Pilotprojekt CARLOS Bericht Nr. 15 Synthese, Auftraggeber: Carlos GmbH, Kommission für Technologie und Innovation, Zürich, http://www.interface-politikstudien.ch/downloads/deutsch/Be_CARLOS_Synthese.pdf [08-02-2010] (2005)

Bähr, J., Jentsch, C., Kuls, W.: Bevölkerungsgeographie, Lehrbuch der Allgemeinen Geographie 9, Berlin, New York (1992)

Barbuto, J.E.: Identifying the sources of motivation in the post-game press conference, an exercise for applying an integrative taxonomy of motivation. J. Behav. Appl. Manage. **4**(1), 41–47 (2002). (Summer/Fall 2002)

Barbuto, J.E., Scholl, R.W.: Motivation sources inventory: Development and validation of new scales to measure an integrative taxonomy of motivation, Psychological Reports, No. 82, pp. 1011–1022 (1998)

Barbuto, J E., Scholl, R.W.: Leader's motivation and leader's perception of follower's motivation as predictors of leader's influence tactics. Psychological Reports, No. 84, pp. 1087–1098 (1999)

BMVBW: BMVBW Verkehr in Zahlen 2005/2006, Vol.34, Editor: Bundesministerium für Verkehr, Bau- und Wohnungswesen, Hamburg (2005)

Bonsall, P.W., Spencer, A.H., Tang, W.-S.: What makes a car-sharer? Transportation **12**(1984), 117–145 (1984)

Castells, M.: The informational city: Information Technology, Economic Restructuring, and the Urban Regional Process. Blackwell, Oxford/Cambridge (1989)

Cervero, R., Griesenbeck, B.: Factors influencing commuting choices in suburban labor markets: a case analysis of pleasanton. Calif., Transp. Res. Nr. **22**(1988), 151–161 (1988)

City of Calgary: Information über das Regional Carpool Program, http://content.calgary.ca/CCA/City+Hall/Business+Units/Transportation+Planning/Transportation+Solutions/Sustainable+Transportation/Regional+Carpool+Programs/Regional+Carpool+Program.htm [18-02-2010] (2008)

Destatis—Statistisches Bundesamt: Im Blickpunkt—Verkehr in Deutschland 2006, Wiesbaden (2006)

V. Handke and H. Jonuschat, *Flexible Ridesharing*,
DOI: 10.1007/978-3-642-11345-1, © Springer-Verlag Berlin Heidelberg 2013

Drosdowski, G., Müller, W., Scholze-Stubenrecht, W., Wermke, M.: Duden, 21. Auflage, Mannheim (1996)

European Environmental Agency (EEA): Variation of CO_2 emissions from transport (EU-27), http://www.eea.europa.eu/data-and-maps/figures/variation-of-co2-emissions-from-2, [24.07.2012] (2012)

Evangelista, B.: Zimride matches drivers, passengers, SFGate 18.08.2011, Online-edition of the San Francisco Chronicle, http://www.sfgate.com/cgi-bin/article.cgi?f=/c/a/2011/08/17/BUP31KO8L9.DTL, [25.08.11] (2011)

Ferguson, E.: The Influence of employer ridesharing programs on employee mode choice. Transp. Nr. **17**(1990), 179–207 (1990)

Gauler, A., Gretz, W.: Entwicklung der Cargo City Süd, Ansiedlungsentwicklung und Beschäftigtenstruktur, HLT-Report Nr. 587, HLT Gesellschaft für Forschung Planung Entwicklung mbH, Wiesbaden (1999)

Handke, V., Jörß, W.: Maßnahmen zur Einhaltung der Emissionshöchstmengen der NEC-Richtlinie—Part 2: Referenzszenario 2000–2020 für Emissionen unter der NEC-Richtlinie (SO_2, NO_x, NMVOC und NH_3), report for the Umweltbundesamt, FKZ 205 42 221, UBA-TEXTE Nr. 36/2007. Dessau-Roßlau: Umweltbundesamt (2007)

Handke, V., Jörß, W., Lambrecht, U., Dünnebeil F: Emissionen und Maßnahmenanalyse Feinstaub 2000-2020, report fort he Umweltbundesamt, FKZ 204 42 202/2, UBA-TEXTE Nr. 38/2007. Dessau- Roßlau: Umweltbundesamt (2007)

Hannula, A.: Dynamic carpool trial in amsterdam—five years later, article from13/05/2010, gasellit blog, insider's view to european software entrepreneurship, http://gasellit.com/2009/05/13/dynamic-carpool-trial-in-amsterdam-five-years-later [22-03-2010] (2009)

Harvey, D.: The Condition of Postmodernity. Blackwell Publishers, Cambridge (1990)

Institut für angewandte Sozialwissenschaft (INFAS)/Deutsches Institut für Wirtschaftsforschung (DIW): Mobilität in Deutschland, Ergebnisbericht des Projekts KONTIV –Kontinuierliche Erhebung zum Verkehrsverhalten, Berlin (2002)

infas/DLR - Deutschen Zentrum für Luft- und Raumfahrt e.V.: Mobilität in Deutschland, Projektpräsentation und Informationsveranstaltung MiD 2008, Bonn/Berlin (2010)

Interview-Blog: "Nach Erscheinung in Heise hatten wir innerhalb von 24 h über 7.000 Auktionen und Anfragen." Im Gespräch mit Walter Demmelhuber (Drijo), interview from 14/02/2008, http://www.interview-blog.de/ebusiness/nach-erscheinung-in-heise-hatten-wir-innerhalb-von-24h-uber-7000-auktionen-und-anfragen-im-gesprach-mit-walter-demmelhuber-drijo [18-02-2010] (2008)

Jörß W., et al. (2010): Emissionen im PAREST Referenzszenario 2005–2020; research report for the Umweltbundesamt, in the context of the PAREST-project: FKZ 206 43 200/01 „Strategien zur Verminderung der Feinstaubbelastung, Berlin (2010)

Jörß, W., Handke, V.: Referenzszenario 2000–2020 für Emissionen unter der NEC-Richtlinie (SO2, NOx, NMVOC und NH3)", project funded by the German Federal Environmental Agency, Berlin (2006)

Katz, D., Kahn, R.L.: The social Psychology of Organizations. Wiley, New York (1978)

Kollmann, T.: Akzeptanz innovativer Nutzungsgüter und -systeme. Konsequenzen für die Einführung von Telekommunikations- und Multimediasystemen, Wiesbaden (1998)

Kollmann, T.: Die Messung der Akzeptanz bei Telekommunikationssystemen. Wissenschaftsjournal 2:68–77

Levin, I.P., Gray, M.J.: Evaluation of Interpersonal Influences in the Formation and Promotion of Carpools. Transp. Res. Rec. Nr. **724**(1979), 35–39 (1979)

Maslow, A.: Motivation and Personality. Harper and Row, New York (1954)

Mühlbauer, P.: Online-Mitfahrzentrale mit Auktionssystem, Telepolis article dated 10-02-2008, http://www.heise.de/newsticker/meldung/Online-Mitfahrzentrale-mit-Auktionssystem-177904.html [18-02-10] (2008)

Müller-Böling, D., Müller, M.: Akzeptanzfaktoren der Bürokommunikation. Oldenbourg. Munich (1986)

Oliphant, M.: The Native slugs of northern virginia, a profile of slugging in the Washington D.C., Masterarbeit at the Virginia Polytechnic Institute and State University dated 12-01-2008, www.commuterpage.com/research/uploads/Oliphant001/Slugging_Report_Oliphant.pdf, [07-03-2010] (2008)

Onishi, N.: Finding a detour to earn a living in indonesian traffic jams, in: The New York Times dated 13-05-2009, p. A6, www.nytimes.com/2009/05/13/world/asia/13indo.html?_r=1 (2009)

Ozanne, L., Mollenkopf, D.: Understanding consumer intentions to carpool: A test of alternative models. In: Proceedings of ANZMAC99, Australian & New Zealand Marketing Academy Conference at the School of Marketing/University of New South Wales in Sydney from 28-11 to 01-12-1999, http://conferences.anzmac.org/ANZMAC1999/Site/O/Ozanne.pdf (1999)

Partzsch, D.: Daseinsgrundfunktionen. In: Handwörterbuch der Raumforschung und Raumordnung, Vol.1. Hannover, pp. 424–430 (1965)

Rogers, E.: Diffusion of Innovations. Free Press, New York (1954)

Ruppert, K., Schaffer, F.: Zur Konzeption der Sozialgeographie, in: Geographische Rundschau, Nr. 6/1969, Braunschweig, pp. 205–214 (1969)

Sharma, R., Sakhuja, N.: Coming before commonwealth games: Mega carpool for Delhi, http://www.thaindian.com/newsportal/business/coming-before-commonwealth-games-mega-carpool-for-delhi_100228465.html [12-03-10] (2009)

Simon, B.: E-Learning an Hochschulen. Gestaltungsräume und Erfolgsfaktoren von Wissensmedien. Cologne: Josef Eul Verlag (2001)

Traffic, Social: Praktijkexperiment. Mobiel en vraag gestuurd vervoersconcept, Eindevaluatie, Haarlem (2005)

Sonnenschein, M., Raabe, T., Reents, G., Hüsken, T., Vogel, U.: Wirkungsanalyse einer optimierten Kommunikationsplattform zur Bündelung des Individualverkehrs mit dem Ziel der CO2-Einsparung, final report, University of Oldenburg, 12-01-2005, http://www-ui.informatik.uni-oldenburg.de/download/CarPooling/downloads/CarPooling-Abschlussbericht.pdf [11.03.10] (2005)

Sparks, P., Shepherd, R.: Self-identity and the theory of planned behavior: Assessing the role of identification with "green consumerism". Soc. Psychol. Q., Nr. **55**(4), 388–399 (1992)

Statistisches Landesamt Hessen: Personenverkehr mit Bussen und Bahnen in Hessen 2007, Wiesbaden (2008)

Statistisches Landesamt Hessen: Bestand an Kraftfahrzeugen mit amtlichem Kennzeichen am 1. Januar 2008, http://www.statistik-hessen.de/themenauswahl/verkehr-umwelt/regionaldaten/bestand-an-kraftfahrzeugen-mit-amtlichem-kennzeichen/index.html (2009)

Taylor, S., Todd, P.: Understanding the determinants of consumer composting behaviour. J. Appl. Soc. Psychol. **27**(7), 602–628 (1997)

Umweltbundesamt (UBA): Emissionsberichterstattung der Bundesrepublik Deutschland 2011, CRF-Tabellen 1990–2009, April 2010 (Stand: 15.04.2011), www.umweltbundesamt.de/emissionen/publikationen.htm (2011)

Umweltsenat - Senat für Frauen, Gesundheit, Jugend, Soziales und Umweltschutz (1998): MOVE-Projekt Report, Abschlussbericht, Bremen, www.move-bremen.org [18-08-2009] (1998)

Urry, J.: Sociology beyond Society, mobilities for the 21st Century. Routledge, London (2000)

Vanasse Hangen Brustlin Inc.: Dynamic Ridesharing Data, Richmond (2006)

Vroom, VH.: Work and motivation. Wiley, New York (1964)

Werlen, B.: Sozialgeographie, Eine Einführung, Bern, Stuttgart, Wien (2000)

Wuppertal Institut für Klima, Umwelt, Energie GmbH: Rahmenbedingungen und Finanzierungsmodelle flexibler Angebotsformen in europäischen Regionen, Projektbericht im Rahmen von IMAGO (Innovative Marketing- und Angebotskonzepte für Gemeinden mit Ortsbussystemen), Wuppertal, http://www.imago-mobil.de/files/IMAGO_BP-Bericht.pdf [08-02-2010] (2004)